D1566205

Geoffroy Saint-Hilaire: A Visionary Naturalist

ÉTIENNE GEOFFROY SAINT-HILAIRE
1772–1844

A Visionary Naturalist

HERVÉ LE GUYADER

TRANSLATED BY MARJORIE GRENE

THE UNIVERSITY OF CHICAGO PRESS
Chicago and London

Dedicated to the memory of Franck Bourdier and Hubert Condamine

HERVÉ LE GUYADER is professor of developmental biology at the Université Paris XI. He is the editor of *L'évolution*. MARJORIE GRENE is professor emerita of philosophy at the University of California, Davis, and adjunct professor of philosophy and science studies at the Virginia Polytechnic Institute.

The University of Chicago Press, Chicago 60637
The University of Chicago Press, Ltd., London

Printed in the United States of America
Original edition: *Geoffroy Saint-Hilaire: Un naturaliste visionnaire*

The University of Chicago Press gratefully acknowledges a subvention from the government of France, through the French Ministry of Culture and Centre National du Livre, in support of the costs of translating this volume.

13 12 11 10 09 08 07 06 05 04   1 2 3 4 5

ISBN: 0-226-47091-1 (cloth)

Library of Congress Cataloging-in-Publication Data

Le Guyader, Hervé.
[Etienne Geoffroy Saint-Hilaire, 1772–1844. English]
Geoffroy Saint-Hilaire : a visionary naturalist / Hervé Le Guyader ;
translated by Marjorie Grene.
p. cm.
Includes bibliographical references (p.).
ISBN 0-226-47091-1 (cloth : alk. paper)
1. Geoffroy Saint-Hilaire, Etienne, 1772–1844. 2. Zoologists—France—Biography.
I. Title.
QL31.G4G8913 2004
590′.92—dc21
[B]
2003053336

∞ The paper used in this publication meets the minimum requirements of the American National Standard for Information Sciences—Permanence of Paper for Printed Library Materials, ANSI Z39.48-1992.

CONTENTS

GEOFFROY S^T. HILAIRE.

INTRODUCTION

The National Museum of Natural History, situated in the Botanic Garden in the fifth arrondissement of Paris, is often considered by its regular visitors to be one of the magic places of the capital. Its privileged location, the majesty of its buildings, the arrangement of the gardens bear melancholy witness to its past greatness.

In the nineteenth century, the Museum was the mecca of the natural sciences. It was there that comparative anatomy and paleontology were founded and codified[1] and that the first theory of evolution was taught.[2] It was there also that cholesterol[3] and radioactivity[4] were discovered. In the course of the first half of the nineteenth century, anything in the whole world that could be of the slightest importance in biology, in mineralogy, and, more generally, in the inquiries of the curious into nature passed through the Museum—hence the extraordinary richness of its collections: minerals, stones, plants, animals. . . . You have to go wandering in the Galleries of Comparative Anatomy and of Paleontology, or else in the recently opened Great Gallery of Evolution,[5] in order to grasp the vast scope of the undertaking of those naturalists who wished to actualize in intelligible fashion "the catalogue of life."[6]

The Museum is a direct heir of the century of Enlightenment. Its foundation by the Convention in 1793 was one of the (numerous) positive elements of those times of fury. From the start, it had as its mission the task of allying research and instruction. In a way that was extremely innovative for the period, it was to pursue studies of a fundamental type, that is to say, studies for which knowledge is a priori the only goal, but without forgetting their applications, as in horticulture or in chemistry. In addition, the instruction there was free and open to all.

Inevitably, for a Jacobin country like France, the circumstances arising with the creation of such an establishment in Paris influenced—for better and for worse—generations of biologists to come. It is there that we must seek the foundations of the experimental biology of a functional cast that was to blossom, thanks to François Magendie,[7] Claude Bernard,[8] or Louis Pasteur,[9] with the success familiar to us. But springing equally from this same source were first the profound resistance to the Darwinian theory of evolution,[10] and then the unpardonable delay in the introduction of genetics as a new biological discipline.[11] To know the first professors at the Museum—the "founding fathers"—is to understand the substance of the biological sciences in France, with their innovations, their wealth, and their inertia.

To learn who they are we can start, for example, by noticing the names of the streets that adjoin the Museum.[12] Let's go together to the Botanic Garden, leaving the Metro at Jussieu—the station named for the celebrated family of botanists that includes Antoine-Laurent de Jussieu, one of the first professors of botany at the Museum and a theorist in botanical classification. Let us take the rue Linné—Linnaeus Street. Here we are honoring the memory of the great eighteenth-century Swedish naturalist, the systematist who was the rival of Buffon (whose street extends along the east side of the Museum). We arrive at a square where one of the principal entrances of the Museum is located. Here three roads terminate: the rue Lacépède, which runs down from Mount Sainte-Geneviève; the rue Cuvier, which runs along the west side of the Museum; and the rue Geoffroy Saint-Hilaire, an extension of the rue Linné. These streets are dedicated to the three zoologists who divided the study of vertebrates at the Museum among themselves at the beginning of the nineteenth century.

What do these names suggest to the average Frenchman? If it is undeniable that many recognize Cuvier, the most famous anatomist of his time, only a few, even among educated people, can place the other two zoologists. What is remembered of Lacépède is chiefly that in 1804 he was the first grand chancellor of the Legion of Honor. In the case of Etienne Geoffroy Saint-Hilaire, there is nothing—or at most he is vaguely remembered as the loser in a certain controversy at the Academy of Sciences.

However, Etienne Geoffroy Saint-Hilaire was a considerable personage. Involved in the founding of comparative anatomy, of experimental embryology, of teratology, and of evolutionary paleontology, he makes his appearance as one of the great French naturalists of all time. Consistently cited by anglophone historians of science,[13] he is as good as ignored in France. This holds even for many biologists, those who consider that the study of the great figures of the past is of little interest for "science-in-the-making." However, this time they are quite evidently mistaken: Geoffroy Saint-Hilaire's unique idea, the idea that was the obsession of his whole life, and yet his continual source of inspiration, gains substance again in the light of molecular biology.

Geoffroy Saint-Hilaire was a man of one idea, which he announced as early as 1796, when he was only twenty-four:[14] "*It seems that nature is confined within certain limits and has formed all living things on only one single plan.*" He was postulating here that all the animals a zoologist could study were constituted in accordance with one and the same organizing principle—this is indeed rather surprising. It is doubtless the mark of genius to have only one idea in a lifetime—but not just any idea. The idea must be simple enough not to be constraining, rich enough to nourish continuing reflection, broad enough to be traversed in a thousand ways. Finally, it must be "of its time"; that is, the scientific techniques necessary to put it to work must be available. This last point, perhaps the most fascinating, illustrates the strange nature of the work of Etienne Geoffroy Saint-Hilaire. Discussed and harshly criticized since 1820, his idea really comes to life only a century and a half later, at the moment when it meets the techniques of molecular biology. This fact invites us to consider this subject more deeply by a rereading of Geoffroy Saint-Hilaire, and by a new look at the celebrated 1830 controversy at the Academy of Sciences that opposed Etienne Geoffroy Saint-Hilaire and Georges Cuvier.

I

TWO FRIENDS OF FORTY YEARS

Etienne Geoffroy Saint-Hilaire, Georges Cuvier: separately one can certainly recognize them, but not understand them. Of the same generation, they were able to traverse happily the social and political upheavals that followed one another in France from the Revolution to the July monarchy. In spite of the circumstances, and with very different characters, the two great zoologists of the Museum both succeeded in remaining constantly productive.

Geoffroy Saint-Hilaire would remain faithful to Bonaparte, who had so fascinated him in Egypt.[1] However, he profited little from that trump card. He twice refused the post of prefect that the emperor proposed for him in order not to desert the world of science. Cuvier, on the other hand, traumatized by the bloody events of the revolutionary era—"I hate the populace," he would say later—would always be on the side of power, whatever it might be. In his youth he had received training both as an administrator and as a naturalist, and, with a hunger amounting to bulimia, he would extend his scientific and political power with every change of regime.

Paradoxically, everything simultaneously united and separated them. Yet history has definitively linked them.

Etienne Geoffroy Saint-Hilaire (1772–1844), Intuitive Genius

Etienne Geoffroy Saint-Hilaire[2] was born April 15, 1772, at Etampes, where his father, Jean-Gérard (1734–1804) was procurator, then judge. A most precocious child, gifted with great imagination, he was quickly destined for the priesthood. Tonsured at twelve, he was a canon at fifteen. In 1788 he entered the College of Navarre in Paris and became a bachelor of law two years later.

He was already following passionately the courses in science, such as those of the botanist Antoine-Laurent de Jussieu (1748–1836).

Revolutionary ideas diverted him from the priesthood, and with his father's consent he turned to the study of medicine. Becoming a student at the College of Cardinal Lemoine, he had as his teachers Charles François Lhomond (1727–1794), a grammarian and an excellent botanist, and René-Just Haüy (1743–1822), latinist and recent founder of modern crystallography. He would also hear the chemist Antoine de Fourcroy (1755–1809) at the Botanic Garden and the zoologist and mineralogist Louis Daubenton (1716–1800) at the Collège de France. He then became the special pupil of Haüy and even had the honor of participating in the informal meetings at which the new crystallography was presented to the great names of the Academy of Sciences, like Louis Lagrange (1736–1813), Antoine Laurent de Lavoisier (1743–1794), Pierre Laplace (1749–1827), Antoine de Fourcroy, Claude Berthollet (1748–1822), or Louis Bernard Guyton de Morveau (1736–1816).

In the revolutionary turmoil, just before the massacres of September 1792, Geoffroy Saint-Hilaire rescued Haüy from his incarceration as a priest. This act would have a capital importance for him, since purely by coincidence it would be the prelude to his prestigious career. In fact, it would be on the warm recommendation of Daubenton, with the vigorous insistence of Haüy, that in May 1793 Jacques Henri Bernardin de Saint-Pierre (1737–1814), then intendant general of the Botanic Garden, would name Etienne Geoffroy Saint-Hilaire subcurator and subdemonstrator at the Cabinet of Natural History.

In June 1793, Joseph Lakanal (1762–1845) reorganized the Botanic Garden and created the Museum of Natural History, following an old project of Georges Buffon (1707–1788), which had been revived by Daubenton in close to its original form.[3] Twelve chairs with the titles of professor-administrator were created.[4] Zoology was at first divided between Geoffroy Saint-Hilaire (quadrupeds, cetaceans, reptiles, birds, and fishes) and Jean-Baptiste de Lamarck (1744–1829) (insects, worms, and microscopic animals). Etienne de Lacépède (1756–1825), who was under suspicion and had had to take refuge in the provinces, returned in July 1793, at the end of the Terror, and was put in charge of reptiles and fishes.[5] Some time later, Jean-Claude Mertrud (1728–1802)[6] obtained the chair of animal anatomy, which would be called the chair of comparative anatomy when Cuvier came to occupy it.

Thus, in the revolutionary euphoria of the time and against the advice—well motivated, we must admit—of Fourcroy, a chair of zoology was assigned to a twenty-one-year-old crystallographer for reasons that had nothing to do with science. But it was a gamble that won. We see the young professor immediately attack his new assignment seriously and enthusiastically. Among other things, when he had barely started, he founded the menagerie of the Botanic Garden, of which he would long be the administrator.[7]

It was at this moment that the Abbé Alexandre Henri Tessier (1741–1837) warmly recommended to him a young zoologist, Georges Cuvier, whom Geoffroy Saint-Hilaire would bring to Paris in 1795. The two young men soon became friends and published four articles together that year, among them two memoirs, one on orangutans and one on elephants.[8] But according to Daubenton, Geoffroy Saint-Hilaire was preparing a formidable rival for himself.

In fact, their thought diverged immediately. In 1796, Geoffroy Saint-Hilaire, attracted by the "world of ideas," took up an idea of Félix Vicq d'Azyr (1748–1794). In a famous memoir on the makis (lemurs of Madagascar), he postulated the existence of a *single plan* for all animals. In 1797, Cuvier, interested only in the "world of facts," posed the problem of classification for the animal kingdom and began to apply the principles that Bernard de Jussieu (1699–1770) and then his nephew Antoine-Laurent had created for botany.[9]

In 1798 we meet a new imponderable that will be important later on. In the greatest secrecy, Berthollet assembled a team of scholars to accompany Bonaparte to Egypt.[10] Cuvier, very probably ill, declined the offer;[11] the ebullient Geoffroy Saint-Hilaire accepted with his usual enthusiasm. During the three years of his stay outside France—he would return to Marseille in November 1801—he brought together the principal ideas that would mark his life as a man of science. Thus, from that moment on, he would employ the principle of a *single plan* to structure all his writings.

During his stay in Egypt, he began to exhibit the nervous disorders that would never leave him. He was subject to serious crises of cyclothymia, no doubt aggravated by the climate, with alternating bouts of intense excitement and great despondency. In the course of one of his euphoric moments he elaborated "the universal law of the attraction of like for like *[soi pour soi]*," which he would take up again thirty years later, at the end of his life. He also erected a grandiose theory of the universe that would earn him sarcastic remarks from the mathematician and physicist Joseph Fourier (1768–1830).[12]

On several occasions during these three years, he would demonstrate his courage and self-sacrifice, which decidedly form a major aspect of his character. Moreover, it is thanks to him that the English, who held the supremacy of the seas after the battle of Aboukir (August 1, 1798), allowed the French scholars and artists to leave with the fruit of their labors, with certain notable exceptions, like the Rosetta stone.[13] Among the collections that Geoffroy Saint-

Hilaire brought back we should mention, in addition to examples of Egyptian fauna, the many animal mummies that would be important later when evolution came to be discussed.

In December 1804, at thirty-two years of age, Geoffroy Saint-Hilaire married Pauline Brière of Mondétour, who was then twenty-four.[14] She was the daughter of a former tax collector under Louis XVI, who, under the Consulate and the Empire, had become mayor of the second arrondissement in Paris. She would bear Geoffroy Saint-Hilaire three children: Isidore in 1805 and the twins Anaïs and Stéphanie in 1809. Later, Isidore would also be a professor at the Museum, and he would chiefly continue his father's work in embryology. In addition, it is he who would invent the term *teratology* to designate the study of the anomalies and monstrosities of living things. He would publish a (good) biography of his father three years after the latter's death.

In 1807, Etienne Geoffroy Saint-Hilaire finally entered the Academy of Sciences, twelve years after Cuvier, who had become one of its permanent secretaries. That same year, Geoffroy Saint-Hilaire went to Portugal to supervise the museum at Lisbon.[15] Instead of broadly pillaging the collections, as the French were doing systematically elsewhere, he taught the Portuguese the art of classifying their specimens, and he proposed exchanges with the Museum in Paris, thus equally enriching both museums. That is why, after Waterloo, the Portuguese were the only people who demanded nothing in reclamation from France. On the contrary, even after the defeat, they again thanked Geoffroy Saint-Hilaire for having taught them to place a proper value on their precious collections.

After his return from Portugal, Geoffroy Saint-Hilaire devoted himself almost exclusively to the investigation of the unity of plan. He had extended it to the totality of tetrapods (mammals, birds, reptiles, batrachians); then, in 1807, in his memoirs on fishes, to all vertebrates.[16] This work is fundamental, for it allowed him to put into shape a method that is really efficacious in comparative anatomy. Moreover, Cuvier was most laudatory, and all seemed to be going well between the two friends.

However, at this moment, Geoffroy Saint-Hilaire, wishing no doubt to extend his principle to other groups of animals, began to change the name of his law; instead of *unity of plan*, he often used the term *unity of composition*. This fact—which would not be definitive until 1818—is important: it allows us to date the moment when, for the first time, the divergence from Cuvier could be suspected in his writing, even if it did not yet openly appear.

In 1810 he was appointed to the chair of zoology at the Faculty of Sciences in Paris, which had been restructured during his trip to Portugal by the imperial decree of March 17, 1808.[17] He courteously tried to cede the post to Lamarck, who declined the offer, invoking his age. The year 1818 was marked by the appearance of the first volume of Geoffroy Saint-Hilaire's major work, the famous *Anatomical Philosophy*, in which he described precisely and in detail his method in comparative anatomy, and which Cuvier himself praised—though not without reservations.

For Geoffroy Saint-Hilaire 1820 is a key year. It was at that moment that he proposed, with characteristic impetuosity, to extend the unity of composition of vertebrates to insects. In fact, at that time it was known, through the work of Marie Jules César Le Lorgue de Savigny (1777–1851) and of Pierre-André Latreille (1762–1833), that insects, and more generally arthropods, followed a single plan. Thus Geoffroy Saint-Hilaire's proposition amounted to uniting two groups, each known to be homogeneous—groups that corresponded to two of the four "embranchements" into which Cuvier had divided the animal kingdom. Cuvier considered this proposition a violent attack. He did not state his disagreement openly, however, but criticized Geoffroy Saint-Hilaire ("underhandedly," according to the latter) among his colleagues at the Academy and the Museum.

From 1811 to 1830, in parallel to his work in comparative anatomy, Geoffroy Saint-Hilaire was equally interested in embryology, and more precisely in what would come to be known as teratology. With the assistance of Antoine Etienne Renaud Augustin Serres (1786–1868), then professor of anatomy in the faculty of medicine, he drew together embryology and classification.[18] In other words, he postulated that an animal of "superior" ranking passes, in the course of its development, through stages that correspond to the adult stages of "inferior" animals.

In this way Geoffroy Saint-Hilaire powerfully reinforced the idea of the unity of the animal world by establishing a connection between the transitory embryonic stages of certain animals and the adult stages of others. Thus he was one of the very first to postulate the *recapitulation*, through embryogenesis, of different degrees of complexity in the animal kingdom. Although in the 1820s the idea of an evolutionary history had not yet been made explicit by Geoffroy Saint-Hilaire, a certain transformism was beginning to have some weight in his thought, something that Cuvier, fixist and creationist, seemed to have understood without being able to accept it.

In 1825, the affair of the crocodiles of Caen erupted. Cuvier had described these bones and, despite obvious differences, classified them among modern gavials. But Geoffroy Saint-Hilaire had been a specialist on these animals since the Egyptian expedition.[19] He carefully examined the fossils and then vigorously contradicted Cuvier, showing how these specimens differed in an important manner from living animals. To classify one of them, he created a new genus, which he presented as an intermediary between reptiles and mammals. Through this antifixist act, he became the founder of what would later be called evolutionary paleontology.

Thus, in a few years, and on two occasions, Geoffroy Saint-Hilaire was led to criticize Cuvier's science down to its very foundations. This controversy reached its climax in 1830 at the Academy of Sciences, on the occasion of the discussion of a communication by Laurencet and Pierre Stanislas Meyranx (1792–1832), presenting an analogy between cephalopods (cuttlefish, squid and octopus) and vertebrates.[20] From this Geoffroy Saint-Hilaire derived the element he had been lacking and was able to unite in a single embranchement vertebrates, arthropods and mollusks. The unity of plan of the animal kingdom thus seemed to be under construction, step by step, and Laurencet and Meyranx were bringing major arguments to bear on the process.

This time Cuvier, out of patience, responded openly. The controversy became public and made an extraordinary stir. The press, particularly *Le Journal des Débats*, *Le Temps*, and *Le National*, published long articles. The great writer Johann Wolfgang von Goethe (1749–1832), who was without question an excellent anatomist, presented the controversy in a German journal.[21]

Stupefied by the public furor aroused by the debates, the two protagonists agreed to stop the discussions, at least temporarily. Geoffroy Saint-Hilaire published the gist of them in his *Principles of Zoological Philosophy*. But the debate resumed, in a different fashion, several months later, around the topic of evolutionary paleontology. For some time, a vaudeville act of musical chairs was performed at the Collège de France. Cuvier defended his position. André Marie Ampère (1755–1836), holder of a chair in physics, heard the arguments, went to report them to Geoffroy Saint-Hilaire, who suggested to him how to reply, and then devoted some time to them in his lectures.[22] But at the time Frédéric Cuvier, Georges's brother, was in the audience. Obviously, he was ready to go! We can imagine the consequence. . . .

Georges Cuvier died in 1832. Geoffroy Saint-Hilaire then tried to popularize his own conceptions. But his last publications were not understood by his colleagues, and the Academy published only the titles. Nevertheless, although attacked by Cuvier's ubiquitous students, he continued to develop his ideas on evolutionary paleontology.[23]

He went blind in 1840 and resigned his chair at the Museum the following year. He died June 19, 1844, at seventy-two years of age. His funeral was magnificent. No fewer than seven eulogies were pronounced. Among others, those heard were: André Marie Constant Duméril, (1774–1840), representing the Academy of Sciences; Eugène Chevreul (1786–1889), then director of the Museum; Jean-Baptiste Dumas (1800–1884), dean of the faculty of sciences; Serres, a second representative of the Academy of Sciences and a long-time collaborator of Geoffroy Saint-Hilaire; and finally, Edgar Quinet (1803–1875), who was present as a friend of the family. According to the report in the *Gazette médicale de Paris* of June 29, 1844: "The cortege was numerous and was composed of scientific, artistic, and literary notables. To be seen there were MM. Arago, Flourens, Duméril, Mathieu, Jouard, Dupin, Elie de Beaumont, Dufresnoy, Rayer, Cauchy, de Blainville, Poncelet, Ballanche, Pariset, David, Victor Hugo, Renaud, Quinet. . . ."

But that didn't stop Cuvier's disciples from burying, for a century and a half, the ideas of Geoffroy Saint-Hilaire.

Georges Cuvier (1769–1832), the "Napoleon of the Intellect"[24]

The difference in temperament between the two men is obvious, especially when we compare the courses of their brilliant careers. While Geoffroy Saint-Hilaire always remained within the world of biology, Cuvier knew how to manage with virtuosity a scientific-administrative *cursus honorum*. From 1795 through 1830, at every change of regime, he succeeded in extending his power and increasing the number of his honors. At the peak of his career he would be simultaneously professor-administrator at the Museum, permanent secretary of the Academy of Sciences, member of the French Academy, inspector general of public instruction, councilor of state, grand officer of the Legion of Honor, baron, and, at the end of his life, peer of France.

Needless to say, such an accumulation of offices and honors did not make him nothing but friends. Thus, Stendhal, who had sat with him at the Council of State, wrote in *The Life of Henry Brulard*, "What was there not of servility and meanness in M. Cuvier's attitude to those in power?"[25] The evidence suggests that this was unfortunately true. This much granted, his childhood is there not to excuse it, but at least to explain it.

Georges[26] Cuvier was born August 23, 1769, at Montbéliard, a town that was then part of the Duchy of Würtemberg. This Protestant region did not become French until 1793, when Cuvier was twenty-four. His father was a former officer of the famous Swiss regiments in France. Thus Cuvier inherited a double Germanic and French culture.[27]

He was a boy of delicate health, gifted with an astonishing memory and precocious intellectual maturity. From the age of twelve he made natural history collections and quickly assimilated the whole work of Buffon. His parents thought of making him a Lutheran pastor, but he did not obtain a scholarship to study theology at Tübingen.

He then changed his orientation, and at fifteen entered the Caroline University, founded near Stuttgart by the duke of Würtemberg and intended for the formation of future staff for the administration of the duchy. Cuvier remained there for four years, until 1788, studying chiefly law, economics, and "administrative practice." But at the same time he followed the courses in botany and zoology, notably those of the anatomist Karl Friedrich von Kielmeyer (1765–1844), one of the cofounders of the German "philosophy of nature." Kielmeyer taught him dissection and technical drawing, an art in which Cuvier would excel. Kielmeyer probably taught him the great unifying theories of the time, like the Scale of Beings of Charles Bonnet (1720–1793),[28] and it seems likely that Cuvier would transmit them to Geoffroy Saint-Hilaire before combating them.

In 1787 he received the gold cross of the "Knights," a distinction that allowed plebeian students to live with children of the nobility. Despite his brilliant scholarly record, when he left the university in 1788, he was not offered a place in the service of the duke. Since he had to earn his living, he left for Caen, in Normandy, as tutor to the son of a family of the Protestant nobility, the Héricys.[29]

For six years he divided his time between Caen in the winter and Fiquainville, near Fécamp, in the summer. He botanized, practiced a little mineralogy, but chiefly dissected and drew the animals at the seashore, in particular marine birds, fishes, and mollusks. Like Linnaeus, he kept a *diarum zoologicum* and a *diarum botanicum*.[30] During these years he exchanged a voluminous correspondence with Christoff (or Christian) Heinrich Pfaff (1773–1852), his friend from the Caroline University, a physician who had remained in Germany. There we already find the premises for the ideas that Cuvier would defend until his death. For example, the wrote to Pfaff in 1788: "I would like everything that experience shows us to be carefully separate from hypotheses. . . . Science should be founded on facts, in spite of systems."[31]

In 1792 he published his first article, dealing with woodlice, which he recognized as small terrestrial crustaceans. At a meeting of the local agronomic society he met Alexandre Henri Tessier, a member of the Academy of Sciences who had fled Paris, where the Reign of Terror was at its height.[32] Cuvier spoke to him of his passions and in particular showed him the drawings of his dissections. Tessier, much impressed, recommended him warmly to Antoine-Laurent de Jussieu and to Geoffroy Saint-Hilaire. Geoffroy Saint-Hilaire took the initiative of inviting him to come to Paris in 1795. He gave him lodgings for a year, and in July had him named Mertrud's substitute in the chair of animal anatomy at the Museum.

In addition to the articles that he signed with Geoffroy Saint-Hilaire, Cuvier on his own published a memoir in which he proposed a new classification of the "animals without backbones." Instead of dividing them into just two classes, insects and worms, he proposed several distinct groups: mollusks, crustaceans, insects, worms, echinoderms, and zoophytes. Lamarck would take up this classification in his class of 1796.[33] In fact, there were many opportunities for Cuvier when he arrived at the Museum; he was in reality the only zoologist, along with the mineralogist Geoffroy Saint-Hilaire and the botanist Lamarck.

From then on, his career went like lightning. We find him professor of zoology at the Central School of the Pantheon.[34] In April 1796, after the reorganization of the Institute (as the Academy of Sciences was also known) by Lakanal, he was named founding member of the new Academy of Sciences, of which, at twenty-six, he was the youngest member. Things sped up after that. In 1799 he inherited Daubenton's chair at the Collège de France, then, in 1802, he succeeded Mertrud at the Museum in the chair of animal anatomy, which he rebaptized comparative anatomy. Finally, in 1803, he became one of the two permanent secretaries of the Academy of Sciences. He was then thirty-four.

Physically, Cuvier was very much of a type;[35] he had great presence, especially in his academician's uniform. His red hair distinguished him. Gifted with a great memory, an easy style of speech, and an excellent hand with a pencil, he was an orator who imposed silence the moment he began to speak. His clarity of exposition at once commanded agreement. A great worker, he at first devoted himself to zoology, until, step by step, administration took over. On the other hand, Geoffroy Saint-Hilaire, who was bald at an early age, had a blundering way of speaking that fatigued his audience, and his swings of mood did him great disservice.

In 1804 Cuvier married Madame Duvaucel, the widow of a tax collector guillotined in 1793, an energetic woman and a very devout Protestant. She had four children from her first marriage, among them Sophie, who was very close to her stepfather and would act as his very intelligent secretary. Four children were born of her second marriage, all of whom Cuvier had the sorrow of surviving.

At the same time he started on an equally booming administrative career. In 1808, under the Empire, he was named councilor of the university. In this capacity he contributed to the organization of the new Sorbonne, where Geoffroy Saint-Hilaire would later be professor of zoology. From 1809 through 1813 he reorganized advanced education in Italy, the Low Countries, and Germany. In 1811, he was made a knight (Chevalier). At the fall of the Empire, Cuvier adroitly negotiated the change of regime. Thus, beginning in 1814, he was councilor of state, and in that capacity, from 1819 until his death, he presided over the interior section of the Council of State in place of the minister.

In addition, he supported the development of the Reformed Church and became director of non-Catholic denominations at the Ministry of the Interior. Thus even their religious choices set Cuvier and Geoffroy Saint-Hilaire in opposition to one another. The latter, destined in his youth for the priesthood, later became wholly agnostic, without, however, sinking into a narrow anticlericalism. His great humanity always came first. Thus, at the moment of the July Revolution, he received in his home the archbishop of Paris, who had taken refuge in the Botanic Garden.[36] As for Cuvier, although he failed to become a pastor, he would remain faithful to his convictions. At that time such opinions had an essential impact on a life in science.[37]

Every day—except Monday, which was reserved for the Institute—from eleven in the morning Cuvier was either at the Council of State or at the Council of Public Instruction. There was no more time to devote to science, and he relied on his assistants for dissection and writing, which he only supervised. And his rise continued: member of the French Academy in 1818, baron in 1819, grand officer of the Legion of Honor in 1824. Some time before his death, Louis-Philippe made him peer of France.

Cuvier died in 1832, in the course of the cholera epidemic that decimated the population of Paris.[38] Unfortunately, he did not have time to finish the written reply he wanted to make to Geoffroy Saint-Hilaire.

His scientific work was prodigious. The collections at the Museum went from three thousand items in 1804 to thirteen thousand at his death in 1832.[39] In zoology his chief publications were the *Tableau élémentaire de l'histoire naturelle* in 1798, then, in collaboration with Constant Duméril (1774–1860) and Georges Louis Duvernoy (1777–1855), the *Leçons d'anatomie comparée* (1800–1805), and finally, starting in 1817, *Le Règne animal*, with the collaboration of Latreille for the insects. The *Histoire des poissons*, which he had begun as early as his stay in Normandy, and which constitutes the basis of modern ichthyology, was written with Achille Valenciennes (1794–1865). It was not published until very late: the first volume appeared in 1828, the ninth in 1832, when Cuvier died, and the twenty-second in 1849.

The principle of correlations constitutes the basis of Cuvier's thought. He stated it a number of times, in particular in the introduction to the *Recherches sur les ossemens fossiles* (*Researches on fossil bones*), which consists of the famous

"Discourse on the revolutions of the surface of the globe, and on the changes they have produced in the animal kingdom," published for the first time in 1812 and republished in 1825: "Happily, comparative anatomy possessed a principle that, when well developed, was capable of making [all the] obstacles vanish. It was that of the correlation of forms in organized beings, by means of which any being could be recognized, at a pinch, from any fragment of any of its parts."[40]

This is in fact a reformulation of Aristotle's principle of final causes, according to which the parts of an animal must be coordinated in such a way that it is comprehensively in harmony with its environment. This functional, physiological type of approach took Cuvier a long way. For him, the relations among different organs are such that in studying one we illuminate the others, since there are in nature possible combinations and impossible combinations. Thus, starting with an organ, we can "calculate" the whole organism:

> The parts of a being all have a reciprocal suitability, there are certain traits of conformation that exclude others; there are others, on the contrary, that necessitate them. Thus when we know such and such traits in a given being, we can calculate those that coexist with them and those that are incompatible.[41]

Cuvier conceives of a living organism in a holistic manner:

> Every organized being forms a whole, a unique and closed system, in which all the parts correspond mutually, and contribute to the same definitive action by a reciprocal reaction. None of its parts can change without the others changing too; and consequently each of them taken separately indicates and gives all the others.[42]

In fact, unlike Geoffroy Saint-Hilaire, Cuvier begins the study of an animal with that of the function of the various organs that compose it. Thus, shortly after his arrival at the Museum, at the beginning of his course, he explained that we must "regard physiology, that is to say, the explanation of animal machines, as the essential and true aim of zoology."[43]

However, when he came to interest himself in classification, he recognized that not everything can be founded on functions. He then adopted the idea of a plan, but, as distinct from Geoffroy Saint-Hilaire, who wanted to see one single plan, he attributed a particular plan to each "embranchement." Thus the animal kingdom was distributed according to four irreconcilable plans, which corresponded to:

four principal forms, four general plans, according to which all animals seem to be modeled and the further divisions of which, whatever names naturalists have bestowed on them, are only slight enough modifications, founded on the development or on the addition of some parts, but which change nothing in the essence of the plan.[44]

We can see why, in spite of everything, Cuvier accepted the program of a Geoffroy Saint-Hilaire—although he was in fundamental disagreement with it—as long as the work was done within the interior of a single embranchement.

Nevertheless, we are far from the strict empiricism within which the paleontologist of Montbéliard often wants to confine himself. For Cuvier, there are two kinds of observation: those that are sources of theory, for instance, those for the elaboration of the "laws of organic economy," and those that fill the place of theory when it is absent. It is not for nothing that he remarked:

I doubt that it would have been guessed—if it had not been learned by observation—that all ruminants would have a cloven hoof, and they alone; or that only in this class would there be horns on the forehead; or that most of them that have pointed canines would be the only ones to lack horns, and so on.[45]

Cuvier's approach? To rely on theory within the limits of its reliability; if theory is uncertain, have recourse to observation. We must recognize that the results of his research were magnificent. In zoology, he was able to combine the principles of classification announced for botany by Antoine-Laurent de Jussieu with a most meticulous comparative anatomy.

We can understand after the fact the glory of the person who brought fossil bones back to life and gave substance to the beginnings of paleontology. In 1831, in *La Peau de chagrin,* Balzac wrote: "Was not Cuvier the greatest poet of our century? . . . Our naturalist immortal reconstructed worlds from whitened bones. . . . He digs up a piece of gypsum, sees a print on it, and cries: Look! Suddenly the stones are animated, death comes to life, the world unfolds."[46]

In 1812 Cuvier published his *Recherches sur les ossemens fossiles,* which laid the foundation for vertebrate paleontology. In his introduction, he defended fixism and thus set up his opposition on principle to Lamarck, who in 1809 had published his *Philosophie zoologique,* which supported transformism. To determine correctly the stratigraphy indispensable for the relative dating of the fossils in the gypsum of Montmartre, he worked in collaboration with Alexandre Brongniart (1770–1847), starting in 1804. From 1808 through 1811 they published together *Géographie minéralogique des environs de Paris* and then, from 1822 through 1825, *Description géologique des environs de Paris*.

In a lecture about his work in comparative anatomy, Cuvier explained, doubtless with a certain false modesty, "I am only a Perugino. . . . I amass the materials for a future great anatomist, and when one appears, I hope he will grant me the merit of having prepared the way for him."[47] Oddly, this wish has been fulfilled in part, but not as he would have wished, for the loveliest flower of his work, the paleontology of vertebrates, would be one of the disciplines that would destroy the concept Cuvier held most dear, that of the fixity of species.

Fixism: that is the error of which he is most frequently accused. But history often preserves only a caricature of the thought of men of science, insisting as if with pleasure on the points in which they have consequently been shown to be mistaken. Granted, Cuvier was a fixist. But we can only admire, on the one hand, the (almost) perfectly objective manner in which he carried out his paleontological work and, on the other hand, the tenacity he displayed in finding the only likely hypothesis that could at the time be opposed to transformism.

The whole of Cuvier's argumentation is founded on the reasoned exploitation of the results of a nascent geology and on the fact that at that time certainty in dating was impossible. While Lamarck, following Buffon, leaned toward the hypothesis of a great sweep of geological time, Cuvier chose that of biblical chronology. All the geological accidents that were described, necessarily of short duration, could be integrated in a "catastrophic" vision of the history of the earth, a vision exaggerated by the author's emphatic style:

> The tearing and upheaval of beds that happened in the earlier catastrophes show that they were as sudden and violent as the last one. . . . Thus life on earth has often been disturbed by terrible events: calamities that initially perhaps shook the entire crust of the earth to a great depth, but which have since become steadily less deep and less general. Living organisms without number have been the victims of these catastrophes. Some were destroyed by deluges; others were left dry when the seabed was suddenly raised.[48]

However, he seems to make a concession to the naturalists who "rely a lot on the thousands of centuries that they pile up with a stroke of the pen," and study "the oldest documents on the forms of animals" that can be given an approximate date, to wit, the animal mummies brought from Egypt by Geoffroy Saint-Hilaire.[49] Dissecting them, he concluded that there were no striking variations, and in particular, after studying the Egyptian ibis, which he interpreted as being a curlew (*Nummenius ibix*), he concluded:

> Thus there is nothing in the known facts that can give the slightest support to the opinion that the new genera that I have discovered or established in the fossil state . . . could have been the root stock of any of the animals of today.[50]

But let us not forget that, to the apparent fixity of the Egyptian forms, Lamarck had a ready reply: since there had been no variation in the environment, there was no reason for variation in the anatomy of the animals.

At that time, the fixity of species was the dominant theory. Several theories were available to explain the results of comparative anatomy and of paleontology, and, within that context, Cuvier's choice cannot be regarded as a fault. The possible theories confronted one another, starting with the affair of the crocodiles of Caen. From 1820 on, Cuvier, absorbed in his administrative work, was much less creative; from then on he did nothing but defend his own ideas, especially against those innovative notions of Geoffroy Saint-Hilaire.

To give a complete account, we must stress the role of precursor that Cuvier has played in the history of science. As permanent secretary of the Academy of Sciences, he had on the one hand to furnish periodic reports on the state of the sciences and on the other to deliver the eulogies of deceased academicians. Moreover, several years before his death he had introduced a course, "History of the Natural Sciences since Their Origin," at the Collège de France. He wanted to present a complete table of the history of his discipline.[51] However, history will remember him as the most celebrated defender of fixism, and also as an anatomist and paleontologist of genius.

The Controversy

Of the same generation, attracted by the same subjects, but with different conceptual approaches, Cuvier and Geoffroy Saint-Hilaire were bound to enter into conflict. This was plain to see, and Daubenton had understood it all since 1795, so striking were the elements that separated them.

For a good number of historians the controversy between the two zoologists belongs definitively to the past. However, in the light of recent findings, we may ask ourselves again about the real meaning of the controversy. Thus it seems appropriate to study it once more with all urgency, trying to examine all its dimensions, including those that have no direct bearing on science.

It is useful first to mention the chief barriers to our reading, those items that hide the Ariadne's thread indispensable to understanding the debate.

Consider, for example, the time of the controversy. If we stick to the title of the work published in 1830 by Etienne Geoffroy Saint-Hilaire, it lasted only a month. In fact the title specifies: *Principes de philosophie zoologique discutés en mars 1830 au sein de l'Académie royale des sciences* (*Principles of zoological philosophy discussed in March 1830 at the Royal Academy of Sciences*). And why not trust the author, who was one of the two protagonists? However, if it is clear that that month of March does indeed correspond to the heart of the controversy, the minutes of the Academy are more explicit: the "hostilities" began at the session of Monday, February 5, 1830, and stopped at the session of Monday, June 7, 1830, with the presentation at the Academy of *Principes*. That is closer to four months.

That "March 1830" has obscured the view of history, while a careful study of the works of Cuvier and Geoffroy Saint-Hilaire in their chronological sequence shows that in reality the controversy lasted neither a month, nor four months, but instead about a dozen years. The public disagreements begin in 1820, and stop only temporarily in 1830. After the July Revolution and the accession of Louis-Philippe, Geoffroy Saint-Hilaire returned to the warpath. The dispute would not be over until the death of Cuvier in 1832.

It is evident, then, that in order to grasp what happened in March 1830, we need to know the vicissitudes that had marked the previous decade. In fact, this is the only way to reply appropriately to the major question of most anglophone historians—the question of whether the debate did or did not have a bearing on evolution. If we stuck to the spring of 1830, the answer would run the risk of being negative: what the protagonists debated was only the plan of organization and comparative anatomy. On the other hand, if we include the years 1820 through 1831, we see immediately that the notion of evolution is the basic source of the controversy. Moreover, as Patrick Tort has subtly emphasized, the title chosen by Geoffroy Saint-Hilaire for his work on the controversy is not neutral: it is clearly there to recall Lamarck's *Philosophie zoologique*, a work published in 1809 by the true founder of transformism, and the object of Cuvier's sarcastic comments.[52] Besides, it was characteristic of the sentimental Geoffroy Saint-Hilaire to invoke the memory of his famous colleague, who had died a year earlier, in 1829.

We may also ask ourselves why the debate at the Academy has remained so famous. Let us not forget that in 1830 French science was flourishing; most of the great scientists of the revolutionary period were still there or had just died. That is what the chemist Dumas, then dean of the faculty of science, would recall in the speech he gave at the funeral of Etienne Geoffroy Saint-Hilaire in 1844:

> Twenty years ago, if a stranger penetrated the sanctuary of the Academy of Sciences, he would have found himself struck to the depths of his soul by a respect amounting to awe at finding assembled around him so many rare geniuses, the pride of France and the shining lights of their century.
>
> Geometricians bowed before Laplace, Ampère, Legendre, Poisson; physicists came to honor Haüy, Berthollet, Chaptal, Vauquelin; and naturalists proudly added to those illustrious and popular names those of Jussieu and of Desfontaines, those of Cuvier and of Geoffroy Saint-Hilaire, long consecrated also by the veneration of Europe.

At the time, Cuvier and Geoffroy Saint-Hilaire were ranked among the greatest anatomists of their day. Their confrontation was bound to excite curiosity. But there was certainly more to it, for the whole of the intelligentsia followed the debate with passion as long articles were devoted to it in the major periodicals of the time.[53]

There seem to be two reasons for such an infatuation. First, we have the very personalities of the two protagonists. The one, Cuvier, in his continual search for honors, seems the very type of the man of power. We have seen that his career appears exemplary in this respect. The other, Geoffroy Saint-Hilaire, is seen as the archetype of the disinterested scholar, avid for freedom. This brief sketch allows us to guess at the political coloration of the participating periodicals—especially in the troubled situation that preceded the July Revolution.

In what is perhaps a deeper way, it appears that the debate transcended the strictly scientific context. This was in fact the period in which it was thought that science could have an important social impact. Some people were already forging the ideas that would later be found in Ernest Renan's *L'Avenir de la science* (*The future of science*).[54] Further, we would see gravitating around Geoffroy Saint-Hilaire, in their turn, Saint-Simonians, Fourierists, and great names like Balzac, Musset, George Sand. . . . For them, the theme of the debate had a bearing on the situation and on the future of man.

2

THE FOUNDATIONAL WORK

The two volumes of the *Anatomical Philosophy* published in 1818 and 1822, are the mature work of Etienne Geoffroy Saint-Hilaire, in which the theoretical tools that had allowed him to revolutionize comparative anatomy are presented and applied. There is no need of texts other than the "preliminary discourses" to allow us to understand the thought and the logic of the great anatomist. Nevertheless, there are certain surprising details. They are all explained by an historical contingency.[1]

The dedications are important as indications of Geoffroy Saint-Hilaire's sensitivity.[2] The first volume, subtitled "On the Respiratory Organs with Respect to the Determination and the Identity of Their Bony Parts," is dedicated: "To the memory of my father, Jean-Gérard Geoffroy, able jurist, upright and courageous magistrate, and of the colonel of engineers Marc Antoine Geoffroy, my brother, who died at Austerlitz." It is well to recall here the force of the ties that existed between Etienne and his brother. In particular, it should be noticed that Marc Antoine participated in the Egyptian campaign as an officer in the corps of engineers, a fact that made yet closer the ties between the two brothers.[3] The second volume, which has as its subtitle "Of Human Monstrosities," carries as its dedication: "To my masters Louis Jean Marie Daubenton and René-Just Haüy, the homage of filial devotion."[4] We find here those who advised him in his youth, who shaped his career, but also who warned him against Cuvier.[5]

The break between the two dedications is striking; first the family, childhood, youth, adventure, then the university system, sponsorships, and tricks of colleagues. From one volume to the next, Geoffroy Saint-Hilaire has left the domain of innocence. Correspondingly, the two preliminary discourses strike us by their dissimilar tones. The first, serene, begins with the announcement of a leading scientific problem—"Can the organization of vertebrate animals be reduced to a uniform type?"—which is discussed technically in what follows. The other, bitter in tone, begins with a recrimination against certain individuals in the scientific community who "adopted those views only with certain restrictions," and then turns to defense and justification.

The result is interesting for the reader, since the two preliminary discourses give a good presentation of Geoffroy Saint-Hilaire's method. In spite of everything, we may ask ourselves what caused this change of tone. To understand it properly, we have to go back to the time before the Egyptian campaign.

It was in 1796 that Geoffroy Saint-Hilaire proposed his principle of unity for the first time in his "Memoir on the Natural Relations of the Makis Lemur L." He wrote:

> A constant truth for the man who has observed a great number of productions of the globe is that there exists between all their parts a great harmony, as well as necessary relations. It seems that nature is confined within certain limits and has formed living beings with only one single plan, essentially the same in principle, but that she has produced variation in a thousand ways in all her accessory parts. If we consider one class of animals in particular, it is there especially that her plan will be most evident: we will find that the divergent forms under which she was pleased to make each species exist all derive from one another. It is sufficient to change some of the proportions of the organs to make them fit for new functions, or to extend or restrict their use. . . . Thus the forms in each class of animal, however they may vary, all result at bottom from organs common to all; nature refuses to make use of new ones. Thus all the most essential differences that affect each family belonging to the same class come solely from another arrangement, from a complication—in a word, from a modification—of these same organs.[6]

This passage shows that at the time the idea of a single plan did not extend very far. Moreover, Geoffroy Saint-Hilaire is perfectly aware of this, since, after his generalizations, he takes refuge inside one class where the plan "will be most evident." Evident in mammals—and, what is more, in primates—it could not be applied to the rest of the animal kingdom; a method to do so was lacking.

But now comes the Egyptian campaign. It will mark Geoffroy Saint-

Hilaire with an indelible stamp, as Darwin will be marked some time later by his voyage on the *Beagle*.[7] The change of situation, the permanence of danger, the novelty of the fauna, will provoke in him a wholesome alchemy. Granted, he will accumulate observations; but, in some obscure manner, he will gradually develop his ideas in total freedom, far from the Museum. Thus, from the time of his return, we will see him give body to his leading principle and methodically extend its field of application.

In 1807 he was still confining himself to vertebrates—don't forget that he held the chair of mammals and birds—; he generalized the principle of unity to the whole of that group. To this end, he compared among themselves the bony heads of mammals, birds, reptiles, and bony fishes. He then established that, against all appearances, they all presented one and the same plan. To demonstrate this, he had the inspired idea of studying the skulls, not in the adult state, but in the different embryonic stages. This led him to observe *the same number of points of ossification* for each animal. In fact, at the end of development, fusions of bones were produced that concealed the original number of parts and prevented the direct reading of the basic plan of the skull in the adult. From that moment, Geoffroy Saint-Hilaire offered us a synthetic exposition of his method:

> Nature constantly uses the same materials, and is ingenious only in varying their forms. As if in fact she were constrained by initial givens, we see her always tending to make the same elements reappear, in the same number, in the same circumstances, and with the same connections. If one organ takes on extraordinary growth, the influence on the neighboring parts can be observed: they then no longer arrive at their usual development. But all are nonetheless conserved, although in a minimal degree that often leaves them without utility: they become, as it were, so many rudiments that testify in some sort to the permanence of the general plan.[8]

In this passage the basic principles that will be discussed in the two preliminary discourses are already present: unity of plan, law of connections, law of balance of organs, importance of rudimentary organs. The theoretical instruments appear appropriate and the demonstrations convincing. At this juncture Cuvier is laudatory, as we can see in his memoir of 1812 on the composition of the osseous head of vertebrates:

> Several years ago our colleague, M. Geoffroy, presented to the class a general work on the composition of the osseous head of the verte-

> brates, only part of which he has as yet published, and which offers ingenious research and most happy results. To explain the multiplicity of bones that are found in the head of the reptile, in that of the fish, and even in that of the young bird, M. Geoffroy imagined taking the head of quadrupeds' fetuses as the basis of comparison . . . and in this way he succeeded in reducing to a common law conformations whose appearance would lead one at first sight to consider them extremely diverse.[9]

However, if we wish to give a universal dimension to his principle, we must move to a higher level through a convincing application at that level. That is the task to which Geoffroy Saint-Hilaire applied himself, the exposition of which is to be found in the first volume of his *Anatomical Philosophy*. Indeed, if the principle of connections is always his guide, it is used here to quite a different power. Now the professed goal of the work is to realize a *philosophy*. In

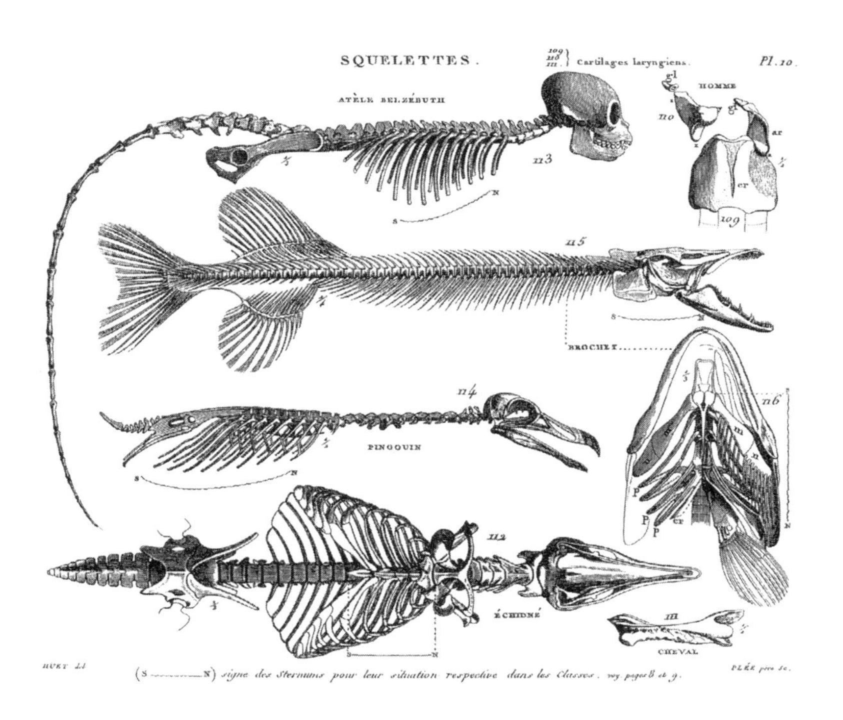

On this plate of volume 1 of his Anatomical Philosophy, *Geoffroy Saint-Hilaire represents together different vertebral skeletons, in order to make plain the concept of the body plan. From top to bottom: a spider monkey, a pike, a penguin, and an echidna (a monotreme mammal). [Top right: human vertebra.]*

other words, where a fragmented observation sees only particularities and diversity, we need first to look for a unity and then to make sense of it.

The preliminary discourse of the first volume is often taken as an example to oppose Geoffroy Saint-Hilaire, "the man of ideas," to Cuvier, "the man of facts." Yet a perusal of the text shows the extent to which this opposition is factitious. Geoffroy Saint-Hilaire tries systematically to bring facts to the support of his theory. On the other side, we know that Cuvier will always have the desire to collect the totality of his observations in a coherent discourse.

In reality, Geoffroy Saint-Hilaire's ambition is never to stop strictly with the work of observation "in abstaining from every abstract proposition." He demands a constant dialogue between theory and observation, but he demands that this dialogue be pushed to the point of becoming *philosophical*, that is, in his eyes, to the point of affecting radically the way a discipline is understood. For him, this must be a *revolution*—in the modern sense of Thomas Kuhn—such that after it the facts can no longer be looked at in the same way.[10]

In the first preliminary discourse we can find a number of traces of this intention. Let us not forget that, from a strictly anatomical point of view, there was very little knowledge of fishes in the zoology of the time.[11] Because of their mode of life they seemed very different from terrestrial vertebrates. It was thus extremely innovative to extend to fishes a proposition that was obvious for tetrapods, that is, chiefly for birds, reptiles, and mammals.[12]

At a deeper level—essential from a philosophical point of view—Geoffroy Saint-Hilaire effects through his method a true Copernican revolution in biology by dismantling a comparative anatomy that had previously been founded on human anatomy, which was taken to be perfect in accordance with the concept of the Scale of Beings.[13] In proposing his principle of the balancement of organs, he demonstrated that the comparative study of a particular organ should rest on a case in which that organ is particularly well developed—which implies that the species referred to as the norm will differ according to the problems posed, and thus will not be Man in every case. However, Geoffroy Saint-Hilaire did not banish the thought of the Scale of Beings entirely from his thought. Here, he was proposing instead a method for running through the scale in a more appropriate manner, without referring every time to its extremity. Thus what we see here is the end of anthropocentrism in comparative anatomy, long after it had been banished from physics. From now on, Man will have no more importance than any other animal; rather, it is through the totality of animals that we try to understand Man.

The second preliminary discourse, corresponding to the presentation of the second volume, seems a logical consequence of the first. We find here, in particular, a defense of comparative anatomy and a new, very elaborate, presentation of Geoffroy Saint-Hilaire's method. However, the bitter tone is astonishing, and we can only be surprised by the theme of the work, human monstrosities.[14] Here, instead of a continuity with the first volume, there is a rupture, which, paradoxically, does not yet appear in the preliminary discourse. This leads us to try to ascertain whether some important event had not occurred between 1818, the year the first volume was published, and 1822, when the second volume appeared.

One glance at the chronology of publication, and the answer leaps to the eye. In 1820 Geoffroy Saint-Hilaire presented to the Academy of Sciences three *Memoirs on the Organization of Insects*. These were published immediately, and then republished in one volume in 1824.[15] We may ask ourselves whether the second preliminary discourse, which was first published independently, should not really be interpreted as a magnificent but abortive act; if it should not rather be considered as material for a work that should have had as its theme "on the relations that insects have with animals said to be unique possessors of a vertebral system"—that is, a restatement of the three memoirs.

After the communications to the Academy in 1820, Geoffroy Saint-Hilaire was sharply attacked, in particular by Cuvier. He felt the need to justify his method once more, and found a new field of application for it that he had not anticipated, namely, the study of monstrosities. He showed that monsters were only variations on a plan of organization identical with that attributed to the normal state. Thus we could classify teratological deviations in a rational manner, as we classify species, by applying the principle of connections. In this way he showed that the same Ariadne's thread was guiding him through all the labyrinths of anatomy.

At this moment the chronology of Geoffroy Saint-Hilaire's work was as follows: 1818, *Anatomical Philosophy*, volume 1; 1820, *Memoirs on the Insects;* 1822, *Anatomical Philosophy*, volume 2, with the two preliminary discourses published later as a whole. The second preliminary discourse is in reality an introduction to and a defense of his work on insects, work to be considered below in chapter 3. Thus we are respecting the internal logic of the sequence.

It was the vast field of application of his theory—from vertebrates to insects, from normality to monstrosity—that led him to change the title of his principle. If the first preliminary discourse begins with the unity of organization, the second concludes with the "Unity of organic composition." Since 1818 the visionary has traversed a very long road.[16]

ANATOMICAL PHILOSOPHY

ON THE RESPIRATORY ORGANS WITH RESPECT TO THE DETERMINATION AND THE IDENTITY OF THEIR BONY PARTS

With drawings of 116 new anatomical preparations

BY M. THE CHEVALIER GEOFFROY-SAINT-HILAIRE

Member of the Institute (Royal Academy of Science); Professor-administrator of the Museum of Natural History at the Royal Garden; Professor of Zoology and of Physiology at the Ecole Normale. Member of the Institute of Egypt. Member of the Academies of Madrid, of Munich, of Göttingen, of Moscow, of Haarlem, of Wettérau at Hanau, of Mainz, of Marseille, of Bordeaux, of Boulogne, etc. And Mayor of Chailly, near Coulommiers.

To the Memory of My Father Jean-Gérard Geoffroy: Able Jurist, Upright and Courageous Magistrate; and of the Colonel of Engineers Marc Antoine Geoffroy, My Brother, Who Died at Austerlitz

1818

Can the organization of vertebrate animals be reduced to a uniform type? Such is the question I propose to delve into in this work.

But, you will say, from what direction could a doubt arise in this regard? Is it not a proposition generally agreed to, and which a natural intuition would have revealed? To return to it once again would be to give the impression that it had not yet been admitted except under the title of prejudice, without examination.

I do not go as far as that, and, on the contrary, I agree that a principle of such universal application must have manifested itself frequently, and even to men wholly foreign to the study of natural history. I may cite an example which is furnished to me by a work that both its importance and the name of its author recommend.

Newton, meditating one day on the simplicity and harmony of the laws that govern the universe, and struck especially by the relations and the uniformity of the masses of the planetary system, abandoned his soul to feelings of vivid admiration: when, suddenly turning his thoughts to animals—those beings whose marvelous organization attests no less, in another area, to the grandeur and the supreme wisdom of the creative power—cried: "*I cannot doubt it; animals are governed by the same kind of uniformity.*"*

This was a first period in the history of analogies: instinct served as guide in the first generalizations. And what shows that they had taken the right path is that naturalists were able to achieve so much more progress in science, the more deeply they were impressed by the correctness of these aperçus.

In fact, it is on the idea that the beings of a given group are linked by the most intimate relations, and are composed of wholly analogous organs, that the scaffolding of the methods of natural history rests—an ingenious art that al-

* Idemque dici possit de uniformitate illa, quae est in corporibus animalium. Habent videlicet animalia pleraque omnia bina latera, dextrum et sinistrum, forma consimili; et in lateribus illis, a posteriore quidem corporis sui parte, pedes binos; ab anteriori autem parte, binos armos, vel pedes, vel alas, humeris affixos; interque humeros collum, in spinam excurrens, cui affixum est caput; in eoque capite binas aures, binos oculos, nasum, os et linguam; similiter posita omnia, in omnibus fere animalibus. Newton, *Optics*, query 31.

lows us to admit as nearly complete the resemblance of a great number of species, although later on we have no more to differentiate them than some slight characteristic traits.

Thus the impetus was given, the categories were prepared, and the goal was indicated. But, it must be admitted, the applications were not always happy; disorder came from a quarter from which there was no reason to expect it. Undoubtedly, naturalists were the first to break the chain they should have continued to use, in order to bring to the unity of composition the most striking diversities. Ariadne's thread escaped their hands, since they did not follow the analogies although they had been clearly discerned from the first approach. Many transformations gave rise to doubts, and from that moment on they were no longer on the same path. Another goal engrossed their minds: it was then a question only of describing and of classifying.

However, things had progressed, and without its being too much the naturalists' fault. Forms are first of all what strikes the senses: they are infinitely various; they seize on our first impressions; they alone occupy us.

Let me clarify this proposition with some examples.

The veterinary anatomist considers the forelimbs of ruminants. In a perfected design he sees a work in which all the parts are in admirable conformity. Does he think of the human arm? What fruit would he reap from that comparison? Wholly contrary to his first impressions, such new forms claim his whole attention; he sees their ends. Perhaps, grasping the relation of these forms with those of every other organic part, or even with the disposition of the places in which ruminants flourish and extend their range, he will even go so far as to raise ideas of harmony. But from then on nothing will turn him from his first impressions. He believes in the existence of new organs; and he has to, since he is creating for himself a new language to paint what he perceives. If he enumerates and describes some parts of this leg, it is of cannon bones, dew claws, hooves, etc., that he informs his hearers, while in ordinary language the same parts are called metacarpals, rudimentary toes, nails, etc.

Who does not see where these consequences lead? It has been observed in itself: people have thought they noticed that the analogies admitted through some vague feeling did not have a character sufficiently determined by evidence. Is one to prefer a principle, even a philosophical principle, to a reality given by observation? From the moment when the question is posed in this way, it is at once resolved: the old traces are abandoned; all views of analogy are thrust aside. A new era begins: we prepare ourselves to found the edifice of science; and since it is believed that there is no possibility of building on a solid foundation except by abstaining from every abstract proposition, we no longer occupy ourselves with anything except tasks of observation.

However, if circumstances made this conduct obligatory, the result was, none the less, that one of the fundamental principles of natural philosophy came to be ignored, that the instruction, the interest, of interrelations was sacrificed to an infatuation with details.

Again, look at what happened to the first monographic naturalists. Wishing to restrict themselves to giving only the characters of species, they settled down in this second era. They adopted the same views; I was almost going to say that they made the same mistakes. Were they in fact concerned with the organ we have just given as an example? The foot of the ruminants became a claw for them in their history of the lion; a hand in that of the monkey, a wing in the description of bats; a fin with respect to the whale, etc. No longer any name in common, the analogy of the parts ceased to be perceived.

But the remedy was there alongside the evil. The multiplicity of those isolated observations led naturalists to collect them, and first of all to look into the relations between them. These tasks were undertaken by naturalists concerned with classifications. The latter, proposing to group entities in order to measure the degrees of resemblance between them, were led to divide into two parts the considerations furnished by each organ, reserving one part for generalities characteristic of the family, and the other for distinctive specialties of particular beings. Thus the Methodologists took the contrary path to the Monographers; they first attached a general idea to the thing in question, in order then to examine its form only in the second place; and thus they found themselves in a position to follow the same organ in all its different modifications.

From this moment on we count a third era: naturalists have returned to the doctrine of analogies. They begin to glimpse this fact of great importance for theory: that an organ, varying in its conformation, often passes from one function to another. For they can follow the front foot just as well in its different uses as in its numerous metamorphoses, and see it successively applied to flight, to swimming, to leaping, to running, etc.; to being here an instrument for digging, there hooks for climbing, elsewhere offensive or defensive weapons; or even becoming, as in our species, the principal organ of touch and consequently one of the most efficacious vehicles of our intellectual faculties.

But how did this return to more reasonable ideas come about? It happened very slowly indeed, and most often without the knowledge of those who were effecting it. To group entities and understand them in a system, in order to refer to it as to a repertoire, was for a long time the principal object of work in natural history.

However, people came to want to know something more than the frame of the picture: they paid more attention to the animals themselves; they compared them with one another and with man. These efforts and new insights insensibly gave a new direction to man's minds: step by step the field of natural history was enriched by philosophical studies, and we finally entered into the present era, which is remarkable for the preference given everywhere to the study of relations.

But, as we can see, the goal was changed without a complete revolution. Attached by habit to the principles of the old system, naturalists did not distance themselves from it except insofar as it suited the circumstances of the moment. Not being in the least worried about the future, they did not know to what point they had made their position equivocal.

In fact, it was quite certain that naturalists had succeeded in attaching a general idea to an organ without introducing into it any notions of its form and its uses. Ask them to define the foot without falling back on those same notions. Astonished at your request, they will reply to you: *this foot, we think of it; that's enough to say.* They will reply to you by invoking authorities, by relying on examples. The ancients had already said *pedes solidi, pedes fissi, pedes bisulci,* when they thought of the appellations "soliped," "fissiped," and "cloven-hoofed." This was imitated by Linnaeus and applied by him as characters to other families: *pedes ambulatorii, pedes gressorii, sensorii, cursorii,* etc.

These authorities are doubtless of great weight; but the more imposing they are and the more they oblige me to avoid leaving the line they have traced, the more they make me want to know the basis of the determinations that have been so positively decreed. I cannot rest content with a vague and confused feeling, and, on the contrary, I am convinced that a practice justified by such constant successes is based on something certain, which it should be possible to erect as a general principle.

Now it is clear that the only generality to apply in the species is given by the position, the relations and the interdependence of the parts, that is to say, by what I include and designate under the name of *connections.* Thus the part of the leg called the hand in man (what is generally known by the word "foot"), is the fourth part of the branch that composes the front member; the terminal portion of that stem, the furthest away from the center of the individual and the most susceptible to variations; the part most especially allocated to communications of the being with all that surrounds it; finally, the stump that comes after the forearm.

It is then that, supported by a precise notion concerning this organ, you see it from above and in its general significance. And you see that from there you can descend, either to follow its different metamorphoses, or to examine its various uses. It is then, I say, that, using all the advantages that such a position affords you, you can return to the phrases characteristic of families, and express yourself in approximately these terms:

"The foot in the bear uses the whole sole or the totality of its bony parts to form the basis of the column that serves to support the trunk. It uses only the metacarpals and the digits in the martens, the digits only in the dog; two in three of the digital phalanges in the lions and the cats; the last of these phalanges in the wild boar; finally, it touches the ground only at one point in the ruminants and the solipeds, dedicating to this not even one part of that last phalanx, but only the nail that encases the extremity."

It is then, let us also add, that we come to rediscover, in relation to the leg (under other forms and with different functions) those of the parts of the sole of the foot that are without contact with the ground in a great number of quadrupeds when they are walking.

At this point in our review we have arrived at a bird's eye view of our subject; we have come to embrace it in what it offers of the greatest generality, and to place ourselves in the most advantageous situation for the comparative study of details. To what are we indebted for this happy position? Obviously, to the principle we have just called attention to, to the principle that saves us from running step by step through all the transformations of organs, and which, when the means of research abandon us, still serves us, and can always serve us, as a guide: to the *principle of connections.*

Such essential services show us the importance of this principle. But in addition, it is also very easy to demonstrate that there is nothing arbitrary in its essence, and that we can perceive in it something quite different from an abstract proposition.

Follow out the idea of this that the very root-stock of the organs can give us. The principal vessels, which are the channels through which (as in the example we proposed) the nourishing fluid is carried to the shoulder, to the arm, and to the forearm, do not stop where the last of these ends. These trees that transport organic seeds extend their branches still further; they must give rise to the last part of the limb and contribute to its support. Such is their destination; there it is, independently of any further result. For in fact it matters little whether the distribution of the molecules of blood takes place in a circumscribed space, or whether this happens in a much elongated line; whether it produces a deposit of which the short, squat paw of the bear is an effect, or whether it favors the elongated conformation of the foot of the deer. The essential point is that each subdivision of the principal branch deposits part of the fluid that it contains and gives its various products exactly, in an order of superposition, which is that of their attachment to the principal branch. Whether or not there is accumulation of these materials, that, I repeat, is a matter indifferent in itself, as long as what can happen only at the base of the column can be referred to its stem.

Such are the organic results, such are the physiological perspectives that can give us an idea of the law of connections and assure us against the fear of seeing its foundation sapped by exceptions: an organ is rather altered, atrophied, annihilated, than transposed.

It was at first by inspiration, and then later, after repeated experiments always followed by success, that I have been making use of this principle for the past ten years (*Ann. du Mus. d'Hist. Nat.;* vol. 10, p. 344). Today I have had to do more, to attempt, by analyzing its essence, to know if it still retained something mysterious, and to show how it happened that I was not misled in according such high confidence to this consideration.

What distinguishes the fourth era, the work of our age, is a well-marked tendency to general propositions, and at the same time a reserve, an extreme circumspection in the employment of means. The end that needed to be reached was then already glimpsed, although still at a remote distance. At first, however, what preoccupied those concerned was the fear of acting with too much precipitation, and they preferred to slacken their pace to make it more fruitfully progressive. It was doubtless possible to arrive in this manner, and we did in fact arrive at greater discoveries, at least at all those which were possible by the methods that had been followed to date.

A new era, whose date is fixed by the publication of this book, begins under other auspices. If it is not a new path that is open, at least the field of organization is illuminated by a new principle, that of *connections:* a principle of high philosophical interest, since it finally admits us to the full and complete enjoyment, without the slightest exception in practice, of that other principle fundamental to natural philosophy, that all animals having a spinal cord lodged in a bony case are made on the same model. The forecast to which this truth leads us, that is, the presentiment that we will always find, in every family, all the organic materials that we have perceived in another: that is what I have included in my work under the designation of *Theory of Analogues.*

Let us try to show that, without the exclusive use of the principle of connections, there arrives a moment in which all the work of determination ceases to be possible. Zoology, for example, catches sight of the relations of all the parts of the front limbs: it can do no more and rests on comparative anatomy to put this proposition beyond doubt. But does the anatomy of animals, to which we owe extremely important work, which has already corrected so many false judgments, and through which we raise ourselves to the most eminently philosophical considerations, truly afford us the means of embracing this problem in all its generality and of giving it a complete solution?

When we examine what has been undertaken in this respect, we are forced to recognize that the usual methods of this science have so far allowed it to grasp and to treat only a part of the question. Let us first follow its course where its procedures have given positive results.

Is it a question of demonstrating that a part of the horse's leg corresponds to the human hand? No one wants to make a direct comparison. But if there exists such a great difference in respect to the conformation of the two organs to be reduced to one and the same type, we flatter ourselves that after all the intermediate degrees have been demonstrated, we will no longer be reluctant to admit the concordance of these parts. So it is that in the last analysis we have to go back to resemblance as conformation, in order to go on to prove the identity of things that are in fact related to one another in many respects, but not in the point being examined.

Supposing that there is nothing that implies a contradiction, there remains, if we are to believe at least in the insufficiency of such a method, there remains, I say, the fear that we could be deprived of some intermediate links—a fear that is by no means exaggerated, since we can apply it to the very case that we have chosen as our one example.

The analogies of the hand have in fact been pursued successfully in air-breathing animals; but when it comes to the fishes, people stop short. In the most remote times, ever since the century of Aristotle, zoology had been inspired by the most happy presentiment and had already related the pectoral fins of fishes to the hands of man. Nevertheless, there was no determination of the bones of the arm and of the shoulder, since no evidence was found of intermediary forms that could lead from one group to the other.

The place where the principle of connections reveals all its importance is especially in the consideration of respiratory systems. Most animals have a larynx, a windpipe, and bronchial tubes; there is none of that in the fishes: so we understand from a comparative study of the forms.

But apply our principle to this observation, and you get quite a different idea of it. The theory of analogues will lead you to suspect that there is no particular and exclusive creation with regard to the respiratory organs of fishes, since in other respects they resemble other vertebrates; and the principle of connections, coming to our assistance, will fortify this presentiment, enrich our research, and finally fix our attention on all the points of a real identity.

Since I had introduced two new means of research into anatomical studies, I found myself caught up in a direction differing in some respects from what had previously been followed. Thus while comparative anatomy makes man its point of departure, and when, relying on the principle that the organs of this privileged species are more perfect, better known, and better defined, it inquires in what and how these organs are diversified, deformed, and altered in all the other animals, my new views lead me not to give preference to any anatomy in particular, but to consider the organs first where they are at the *maximum* of their development, in order then to follow them step by step to the zero of their existence. In the first case, that in which man is placed at the center of a circle, the movement is by a large number of paths or different radii to all the points of the circumference. I move, on the contrary, from that circumference to the center. I confront directly the most shocking anomalies, so as to embrace them in one and the same thought, and to make it plain to see that all these very diverse types of organization converge on the same trunk, and are only more or less differing branches of it.

I shall not stop to consider the physiological consequences of this proposition; it is for the body of the work to offer them. But there are other consequences of its habitual application that it is close to my heart to insist on.

In fact, if it is easy to reduce to the unity of composition the diverse forms of organization of the vertebrates, nothing prevents our young people from holding to a very small number of considerations in their studies. With the principle of connections, you no longer have to fear that the intermediate links will be missing. On the contrary, you are in a favorable position, in a position to do real research, since you can restrict the field of observation only by limiting the number of your examples, and since you can profit from them only by choosing them at great intervals to one another.

Strictly speaking, it will suffice to consider man, a ruminant, a bird, and a bony fish. Dare to compare them directly, and you will arrive in one leap at all that anatomy can provide for you that is most general and most philosophical.

Otherwise, if you continue to run through all the intermediate links, you are embarking on a long and painful voyage. How many people, to whom it would have been both useful and agreeable to undertake this, have been obliged to give it up because they could not devote the necessary time to it? In the same way overseas voyages were available to only a very small number of men as long as we were without the compass and had to follow the coast.

The principle of connections, like another compass, thus brings together the different points of the theater of our explorations. In simplifying our investigations, it puts philosophical anatomy at the disposal of a greater number.

I flatter myself that this new method of research will one day have some influence on medical studies. It will probably deliver young students from the painful uncertainties they feel. For if they desire keenly not to be held to the anatomy of a single species, to a purely surgical anatomy, they fear even more to embark on an enterprise they believe to be beyond their powers.

In concluding my work in this way, let me be permitted to add that I shall regard myself completely repaid for my work, if my research should one day exert such an influence. Yes. Could I but learn that they have been of use to the youth of our schools! What class of our fair France is more worthy of interest? What devotion, what application, what ardor for study! Lovable youth, wholly preoccupied with the noble productions of the mind, you seem to be absorbed in one single thought: in the thought that made Virgil say:

Felix qui potuit rerum cognoscere causas!

ANATOMICAL PHILOSOPHY

OF HUMAN MONSTROSITIES

Work Containing a Classification of Monsters; Description and Comparison of the Principal Genera; a Reasoned History of the Phenomena of Monstrosity and of the Primitive Facts That Produce It; New Views Concerning the Nutrition of the Fetus and Other Circumstances of Its Development; and the Determination of the Different Parts of the Sexual Organ, in Order to Demonstrate the Unity of Composition, Not Only in Monsters, Where the Alteration of Forms Makes This Organ Unrecognizable, but in the Two Sexes, and, Further, in Birds and Mammals.

With figures of anatomical details

BY M. THE CHEVALIER GEOFFROY SAINT-HILAIRE,

Member of the Royal Academy of Sciences; Professor-administrator of the Museum of Natural History at the Royal Garden;
Professor of Zoology and Physiology at the Faculty of Science.
Of the Institute of Egypt. Free Associate of the Royal Academy of Medicine.
And of Several Other Academies, National and Foreign.

To My Masters Louis Jean Marie Daubenton and René-Just Haüy, the Homage of Filial Devotion

1822

I thought I had sufficiently established, in the preliminary discourse of the first volume, that the relation of beings, the analogy of their organs, and the invariable connections of their parts, were necessary effects. Hence I was much astonished to learn that excellent minds, even among those scholars who honored me with great good will, adopted these views only with certain restrictions. That my ideas should have been rejected in their totality: that would have surprised me much less, or even not at all.

Thus the celebrated Dr. Leach* announces in England that my *Anatomical Philosophy* is a prime example worth imitating, in that, he says, it "has opened a wide and new field, and, if followed up will ultimately lead to a more complete knowledge of true comparative anatomy"; and Professor Friedrich Meckel,† in Germany, considers the general propositions of my work so evident, that he supposes they have been imagined for a long time,‡ and that he believes they are adopted by the majority of anatomists.

Nevertheless, both these scholars shortly seem to give way to another chain of thought: they allow themselves to be surprised by some details, in which they finally find the character of serious objections. The former revises some of my determinations while still retaining my nomenclature, whose meaning he then has to change; and the latter calls the principle of connections "a *law* that follows nature with a *pedantic* affectation," and remarks almost immediately that this law is not followed in a great number of cases. I have devoted the paragraph on page 434 to a discussion of these contradictory ideas.

If the order of the universe does not hold to a concatenation of causes and effects, if it is not necessary to consider the animals distributed over the globe as isolated parts of one another, we shall not have far to go to slip backward, and to return to the old way of studying natural history. It is not yet thirty years since zoology was made to consist in the observation of certain parts, like teeth, digits, fin rays, tarsal bony pieces, etc.: parts privileged through the exclusive attention of which they were the subject. Relations were not admitted except

* *Comparative Anatomy*, Annals of Philosophy by Th. Thompson, D.M., no. 92 (1820), 102.

† In the preface of his new *Treatise on Comparative Anatomy*, published last year.

‡ On the priority of these ideas, see the note on p. 445.

for exactly those required to establish a good specific character. For what was proposed was to introduce into the great catalogue of beings the animals newly discovered. And in fact everything appeared to be said about them if their name had been found and a descriptive or characteristic phrase well composed for them.

What less does a librarian do, who restricts himself to judging the format and reading the frontispiece of a new book that has been sent to him? He then knows enough about it to set this new production in the place called for by his system of classification.

The librarian who arranges his books and the naturalist who classifies his animals are at the same juncture: in vain they have to repeat the same actions for every novelty they receive; they never learn any more about the foundation of things. But nevertheless, the philosophical history of the conceptions of the human mind will be revealed to the former, as the philosophical history of the phenomena of organization will be to the latter, if the librarian is at the same time an informed and judicious student of literature, or if the naturalist is equally a physiologist who has both seen much and made many comparisons.

To admit the continual return of the same parts, to the point of seeing in it a formal tendency or a *law* of nature, and then to show that this is not the case, proving it by many references: that is to place oneself between the old and the new school. It is to stop in midvoyage. Thus many investigations had given you a full conviction of the reality of that law, and you falsify it for some considerations that lead you to doubt. But take care! This is not at all to demonstrate good faith and prudence: it is solely to admit that you renounce any philosophy of science. In that case, do not speak of the law or of general facts: act as you used to do, and stick to the observation of isolated facts.

Further they say: "I reject such and such a determination, and I replace it by such another one." Is the case, then, that we can make decisions in the sciences for reasons of convenience? I understand that one can be in total disagreement about the object of a single consideration in research—on the shape of clouds or in the contemplation of things as indecisive and as fugitive as that. But can it be the same in our determination of organs? And if x is to be sought, can it be interpreted indifferently by a or by b? I affirm that our immortal Buffon was born at Montbard on September 7th, 1707. Will you be permitted to confront me with your disagreement, while you investigate the question whether a different date would not be more suitable? Before you try that, start by proving that I am mistaken.

Doubtless one can always choose between different sides; but there is also the risk of grasping the false in place of the true. Thus, M. Magendie tinkers with a phrase (*Journal de Physiologie*, vol. 2, page 127), and believes he is overturning my doctrine on the analogy of organs.

Well! When you succeed in encountering together several animals of one and the same class, like a horse, a cat, a dog, etc., if you cannot consider them without protecting yourself against the impression of the analogy of their parts; if every sense organ, those of locomotion, all the others, in fact, exist in all the animals, are obvious in the same way in all the forms, acting in the same way; if it is not a distinct object that responds in the one as in the other to the appeal you would like to make to it; and if, giving in to a sort of instinct, to an inspiration that does not rely at all on motives in science, you cannot escape the necessity of calling by the same name so many corresponding parts, would you hesitate to believe in the same identity of the internal parts? You will hesitate when you have to acknowledge that the latter are nevertheless only the roots of the former; that the one set continues in the others, and that it is by the same devices that all those parts so manifestly similar externally operate within?

I am anxious to conclude only for fear of insulting the sagacity of the reader. He could not doubt that the study of organization rests on fundamental rules. Nothing arbitrary can be introduced here; and our unknown x will necessarily be a or b, the one to the exclusion of the other, one of the two without the least hesitation.

I have long been occupied in seeking the principles for these rules; and if I have finally gained confidence in some of their applications, if I have been seen, thanks to their assistance, to draw certain conclusions; and if I have given a form to this work through appellations whose novelty of expression was made necessary by the novelty of the objects to be made known, I did not believe it was excessive to use a volume to discuss my motives.

In writing this, it is not that I dream of protecting myself against any criticism. I myself see too many difficulties there. How can there be agreement on the conclusions if there is disagreement at the start about principles? It is only too ordinary to be judged on new views that have not yet been valued by the feelings still retained for the old ones, to be judged by any one who allows himself to be surprised through preconceptions of his own superiority, and by the heedless remark, "Well, I am thinking, and I have always thought, differently."

For it is to this that attention is seldom given: that the principles of a science change successively as the meanings of the words change that are used to hallow its aphorisms. Such a revolution was certainly inevitable in a science as little advanced as general anatomy. It was no use wanting to hold with fidelity, with all possible rigor, to the exact value of expressions of use in one's own time. Originality creeps in in spite of oneself, since, for every little bit that science is made to advance, that is, has the generality of its ideas extended, on the same ground there is an extension of the bearing of its terms, which a more restricted need had produced with the character of an earlier age.

The hesitation of many minds with regard to our present position would thus result from its novelty. To understand how this position is an effect of time, and depends on the progressive order of ideas, let us look at the origin of things. We want to know why and how people resorted to anatomy.

Anatomy, as I understand it, and as I think it will be understood one day—I mean to say, anatomy in all its generality, seems to me to have assumed up to the present three fairly distinct characters, and then I could also add that these three principal modifications are at the same time related to three successive periods. Philosophical among the Greeks, zoological in our day, and entirely medical shortly after the revival of letters in Europe, general anatomy was at first of interest only as completing the one anatomy wished for at the time: it was not appealed to, it was not consulted, except to illuminate certain obscure points of human anatomy.

This division of anatomy, far from presenting the parts as independent of one another, shows us the latter, on the contrary, under the aspect of three branches sprung from a single trunk, three schools produced by one and the same thought, by a conviction that preceded the era of science, in fact, by the presentiment that all beings are formed on one and the same pattern, modified only in some of its parts.

Further, as anatomy was imagined from its origin, so did it remain among the Greeks. There it retained its philosophical character, its condition of generality, since, being closer to its cradle, and consequently constantly attached to the system of organic uniformity, it did not give occasion to any supposition of a different anatomy, *human, veterinary, and comparative.*

To all intents and purposes, the school of Aristotle knew only one anatomy, general anatomy. This did not prevent it from establishing with great sagacity the differences between many particular forms of organization, just as it can be said that we likewise admit only one single zoology, that is, general zoology. Nor does this prevent us from presenting the tableau of classes and of families, for which it is no more difficult to specify the differences as well, since to arrive at the facts that characterize each group or each animal in particular, it is necessary only to descend from the height of the most general considerations.

Any one who would think today of saying "comparative zoology" would arouse indignation, for this would appear to be a totally insignificant pleonasm. If this is so, "comparative anatomy" will not be spoken of for much longer either. To ideas of the same order belong appropriate terms. In fact, zoology is the description of the external organs of animals, as anatomy is that of their internal organs.

This conclusion is rigorous. For I regard as without value the objection that could be raised, that zoology extends further to other considerations, since if one sets aside the visibly perceptible features of the organs, in order to hold to the final object, that is to the usage or the activity of the organization, I mean, to remain fixed on the apparatuses for action, and on all the other manifestations of life, it is true that one arrives at the secondary attributes of zoology. One is proceeding to the second section of the science, treated from so high above and with all the richness of the most harmonious style in NATURAL HISTORY—on the customs and habits of animals. But at this moment we find ourselves at the same point as in anatomy, after descriptive considerations have been exhausted. For there are also the actions of their organs to be narrated, which is expressed by the word "function." Thus in the two cases you are considering the form and the activity of the organs: the form, which is properly the subject of anatomy and of descriptive zoology, and the activity of the organs, as one and the same thing, even though you call it functions or habits: expressions which, whatever you may do, do not differ at all in their application, or rather, which become synonymous. Thus zoology would have separately its physiology as well as its anatomy; no difference in the results, but only in the terms.

These are very natural deductions from what precedes. However, there is nothing astonishing in the fact that they have not been presented sooner: they arrive at their appointed hour, like all that depends on the interconnection of ideas.

And, indeed, such is the character of our time, that it becomes impossible today to restrict oneself severely within the framework of a simple monograph. Study an isolated object, you can relate it only to itself, and consequently you would have only an imperfect knowledge of it. But look at it in the context of beings that approach it in several respects and which are more distant from it in others, and you will discover more extensive relations. First, you will know it better even in its special character. But further, considering it in the center of its sphere of activity, you will know how it behaves in its external world, and all the qualities that it itself receives from the reaction of the surrounding milieu.

People have been comfortable with the path they have followed up to the present, from the previous observation of facts. But in the progressive order of our ideas, it is now the turn of philosophical investigations, which are nothing but the concentrated observation of the same facts, that observation extended to their relations and generalized through the discovery of their connections.

Zoology, which builds up its wealth of the knowledge of the diversified forms under which life reproduces itself, does not really exist except through comparisons. It is thus necessarily *comparative*, as anatomy dare not and cannot ever cease to be, as long as anatomy holds only to a single character, to being only an organic topography. In that case I see there no more than one of the branches of science, one part applicable only to one of the arts of society, doubtless to one of the most important, since it is only on it that most hygienic knowledge can be founded. Such is the part of anatomy even more necessary to the surgeon than to the physician.

Not that it was after these reflections, which would have demanded more study and more maturity than were then possible, that, at the revival of learning in Europe, medical zootomy was nevertheless seen to enter into the philosophical paths of analogy just as plainly as anatomy had done among the Greeks. Only one thought occupied all minds at that moment, that of procuring for physiology better and better assured foundations. But since nothing beyond that had been imagined, people were indeed forced to hold to the doctrine of the Greeks. A proper distrust of their own powers inspired this conduct in every one. And this lasted so long that, for lack of a good exploratory method, it appeared that there was greater advantage in seeking out the materials of science in the masterpieces of the Ancients, where they were found to be elaborated, than in the works of Nature, where it had not yet been learned how to discover them.

It does not enter into my subject matter to examine how this course of events necessarily directed the century of erudition (in anatomy in particular, as in all that belonged to the domain of intellect). What it suffices for me to remark is that medical zootomy found itself at that moment in a position to guarantee it many reefs, the dangers of which would later be recognized.

This danger was occasioned by the multiplicity of investigations and by the different directions taken by the human mind.

The first scholars had harvested in the field of grammatical remarks. Those of the following age gave to their interpretations the authority of the actual observation of the objects. These accessory investigations opened a new path, and that path was followed almost at once for its own sake, to the point that it was forgotten how it had been entered on. Before long the question was no less than the entire reconstruction of the edifice of the sciences. It was then that the study of particular facts began. The Greeks had descended from the interrelations of these facts to the consideration of their differentiating characters. Thus the method of the moderns was the inverse of that of the ancients.

Anatomy, philosophical among the Greeks, restricted itself in the past century to being monographic. It was bent to suit all our needs, and it became *human* anatomy and *veterinary* anatomy to the profit of the two principal species on the interest of which our social economy is founded.

Perrault had understood it in the same sense as the Greeks; and to reduce it to is primitive character, to generality, he had conceived the idea of monographic anatomies of animals, the collection of which was placed at the head of the Memoirs of the Academy of Sciences. This could only be, and was only, in the thought of that great academician, material for a general anatomy. Nevertheless, the gathering of these monographs, where there were only facts to be compared eventually, was again considered a third sort of anatomy, under the title *comparative anatomy.*

In fact, this third sort assumed a wholly zoological character, when, enriched by the genius of Campers, Pallases, and Cuviers, it was so ably and so happily employed in the philosophical investigation of the natural relations between beings.

It is in these circumstances that I published the first volume of my *Anatomical Philosophy.*

I had had some reason to believe that the new views of that work would not have obtained the sanction of the most illustrious of our anatomists. I desired a public explication; I even solicited it in my writings on the insects. What was my satisfaction, when, on February 19, 1821, I heard Baron Cuvier, in a report to the Academy of Sciences,* express himself on the new determinations of the organs as I would have wished to be able to do it myself. I saw that we differ only in expression, happier, firmer, and more elevated in the case of my learned colleague. These ideas are complementary to those that I desired to present in this discourse; I here submit the text.

> Any one who has taken the trouble to approach a certain number of natural beings of the same kingdom or the same class, must have perceived that in the midst of the innumerable diversities of

* See the *Annales générales d'anatomie comparée*, published at Brussels, vol. 7, p. 397.

form and color that they present, certain connections prevail in the structure, the position, and the respective function of the parts, and that with a little attention we can follow these connections through the differences that sometimes conceal them from superficial observation.

A study at a little greater depth even shows that there exists a kind of general plan that can be followed for a longer or a shorter time in the series of beings, and of which we can find some traces in those that were thought most abnormal.

In fact, we have come to recognize that their very diversities are not thrown haphazardly among beings, but that those of each part are linked to those of other parts according to certain laws, and that the nature and destination of each being in the totality of this world are determined by the combination of the diversities that characterize it.

These resemblances, these differences, and the laws of their combinations form the object of the special science to which has been given the name of comparative anatomy, a very important branch of the general science of organization and of life, essential base of all particular natural history of organized beings.

One of the greatest geniuses of antiquity, Aristotle, was the creator of this science, since he was the first to envisage it from this lofty point of view. But immediately after him, the kind of research that could extend his ideas was wholly neglected. And after the renewal of the sciences, attention was paid for a long time, and with good reason, to partial observations rather than to general mediations.

The philosophical spirit which in our day has brought light in most of the sciences of observation has restored comparative anatomy to its dignity, and has once more made of it the regulator of zoology. For some years great movement has also been noticed with respect to it. The most precious observations are gathered, the most delicate connections are grasped. Everything that has been discovered unexpectedly, and in a way miraculously, has seemed to justify the greatest daring in our conceptions. These have gone, so to speak, to recklessness. And already we have seen philosophers wanting not only to link all animate beings together by successive analogies, but to deduce a priori the general and particular composition of the universal laws of ontology and of the most abstruse metaphysics. Any one who has devoted a bit of study to the history of the human mind, without sharing all the views of these authors, will nevertheless congratulate the natural sciences on them. Many men would not enter into so painful a course, did not great hopes excite their ardor.

It is easy to predict, and experience already proves that good fruits will necessarily result from this. Even if their authors did not achieve their aim, they would always have collected on the way an infinity of facts and of views that would nonetheless be solid treasures for science.

Thus, up to now, no one can doubt that the skull of vertebrates is reduced to a uniform structure and that the laws of its variations are determined, etc.

Such is the exposition of the most recent efforts of the human mind on the subject matter of our usual meditations, made from above, and as it behooved a great talent to present it.

However, what belongs to us in this general movement of minds? We shall say it without affecting false modesty. The intention was not to move away from the Aristotelian path, but it lacked rules that could guide us on the voyage. It was at this juncture that I made known a *New Method* for arriving more directly and with greater certainty than had formerly been possible at a real *determination of the organs.*

This method, a true instrument of discovery, consists in the intimate association of four rules or principles, the definition of which I have summarized under the following designations:

The theory of analogues, the principle of connections, the elective affinities of organic elements, the balancement of organs.

1. The first of these principles forms the basis of the doctrine of Aristotle, but, resting less on a demonstration than on a feeling, it was bound to be, and was, usually abandoned in practice. In fact it was necessary to confine oneself very strictly to the consideration of beings of a given Class, or more precisely, to that of beings of a given Order, if one did not wish to arrive at numerous exceptions from all sides that would destroy the universality of the rule. Otherwise, would anyone ever have thought of discovering a veterinary anatomy distinct from human anatomy? But I have regenerated that principle and have procured for it an omnipotence of application, by demonstrating that it is not

always the organs in their totality, but instead only the materials of which each organ is composed, that maintain their identity. This is then understood in such a way that the philosophical thought of the analogy of organization constitutes my first rule, that is to say, the *theory of analogues.*

2. But further, I have given this rule a necessary support, without which, in fact, the theory of analogues would have seemed to be just a glimpse by the mind, that is, the *principle of connections.* Formerly there was talk of analogy, without anyone knowing what in particular was analogous. For want of anything better, there was endless talk about the consideration of forms, but no one appeared to see that the form is fugitive from one animal to another. Thus I will have provided the consideration of analogy with a basis it had previously lacked, when I proposed to bring research *uniquely* (p. 447) to bear on the mutual, necessary, and consequently invariable, dependence of the parts.

3. The materials of organization are grouped together to form an organ, as houses agglomerate to compose a city. But divide that city, as has been done in Paris, into several municipal governments, and it will not be arbitrarily, but always through a necessity of position, that the dwellings, or our organic materials, will be distributed. This necessity, which constrains adjacent elements into accepting the effects of a reciprocal suitability, is what I understand by the *elective affinity of organic elements.* For further details, see the paragraph on page 387.

4. Finally, I call balancement between the volumes of organic masses, and in abbreviated form, *balancement of organs,* that law of living nature, in virtue of which a normal or pathological organ never flourishes to an extraordinary degree, without its being the case that another organ of its system or of its relations suffers to the same degree. I often return to this idea, but on page 244 I have made it the subject of particular reflections.

I cannot doubt the practical utility of these four rules. I have tested them even on subjects where I thought their investigative power would stop short. In other words, when I was seeking, through them, to take account of the most disparate facts of regular organization, of the connection that insects have with animals that are said to be uniquely in possession of a vertebral system, or when I proceeded to study, in the facts of monstrosity, the most reckless and disordered organization.

But it was by no means to partial successes that the *new method* had to be limited. For, whatever system of organization it is applied to, and, in general, to whatever point its action is directed, it gives identical results. It tends to reproduce as a fact acquired a posteriori the guiding and fundamental idea of the philosophy of Leibniz, the idea which that great genius summarized in the expression, *variety in unity*.

This general and definitive result of my determinations of organs has become the most elevated conclusion of my investigations, high manifestation of the essence of things, which I have expressed and proclaimed under the name of *Unity of organic composition*.

3

THE PROVOCATIVE WORK

The work *Mémoires sur l'organisation des insectes* (*Memoirs on the organization of the insects)*—which has been examined here on the basis of the publication of 1824[1]—brings together the three reports that Geoffroy Saint-Hilaire presented to the Academy of Sciences on January 3 and 18 and February 12, 1820.[2] Their three subtitles are explicit about their intended purpose. In fact we discover in succession:

- On a skeleton in the insects, all the parts of which, identical among themselves in the different orders of the entomological system, correspond to each of the bones of the skeleton in the higher classes. (By higher classes, it is appropriate to understand the classes of vertebrates.)
- On certain fundamental rules in natural philosophy.
- On a vertebral column and its ribs in the "apiropod" insects.

If we take it literally, the reasoning Geoffroy Saint-Hilaire follows in these memoirs is baffling enough at first sight, and it is easy to understand Cuvier's anger. In fact, after he has explained the general organization of insects and has remarked "that we find in the insects, contained at one and the same time in a single tube, not only their spinal cord, but all their abdominal organs," he draws from this conclusions that are, to say the least, unreasonable:

> From these facts it is to be concluded that insects are vertebrates; and if everything can be reduced to a vertebra, it is in the insects that this proposition appears in all its clarity. In the last analysis, we arrive at this result: every animal lives within or outside its vertebral column.

We have in fact from now on this great characteristic to differentiate the old vertebrates from the new, who will take their position after the others.

We must be very clear here. Following the comparative anatomy of the time, including that of Geoffroy Saint-Hilaire, the reasoning is mistaken. The error consists in giving the name "vertebral column" to the cuticle that, in the case of the insects, produces an external skeleton. But as the embryology of the insects is very different from that of the vertebrates, it is entirely impossible to apply Geoffroy Saint-Hilaire's principles—in particular, the principle of connections—and to arrive in a convincing manner at objective considerations as striking as those he had used in the first volume of his *Anatomical Philosophy*. Breaking with his own innovations, he omits the study of the embryological stages and compares only the adult stages. So he lets himself be guided by his intuition, which we know to be galloping. The danger is there and Geoffroy Saint-Hilaire knows it. Indeed, he writes in this memoir: "Analogy is advisory: let us reflect on this well, for fear of its consequences or its abuse." But perhaps the limits are difficult to estimate, as, indeed, he declares as early as the introduction to the first volume of the *Anatomical Philosophy:* "Between these two extremes"—i.e. to make up one's mind in terms of analogy alone, or to stick too stubbornly to the facts—"it seems to me there is a middle course to take." Here Geoffroy Saint-Hilaire finds himself in an untenable position. While he has views of wide import on *composition*, he has no instrument with which to study it; so he tries to reason on the ground of *plans of organization*. Hence the difficulties in terminology and the difficulties in logic.

There are now two possible readings. The first is to admit Geoffroy Saint-Hilaire's error and reject the totality of his conclusions without studying them. That is what Cuvier did. But it is also possible to consider that he had a brilliant intuition, and that he was able to assent to it in complete confidence even through the skewed perspective of his error of reasoning, which was in this case beneficent.

Still, we can understand how Geoffroy Saint-Hilaire was tempted. At this period, as he tells us, Savigny had already worked out the comparative anatomy of the insects—in particular with respect to their appendices and mouth parts—and had demonstrated the identity of plan in the latter. Latreille had extended these ideas to crustaceans. Thus, within each of the embranchements that had been chiefly investigated, that is to say, the vertebrates and the arthropods, work had progressed in the same fashion, and the principle of a single plan had been applied with pertinence and profit. Finding a continuity between the plans of organization of the insects and the vertebrates permitted a move toward a global unity at a higher level, that of the animal world.

But Geoffroy Saint-Hilaire went still further than that. He wanted to find a homology between the different parts of the organisms of the two embranchements:

> The first segment of the body of the insect does not correspond to the whole head of the vertebrate, but is composed of the bones of the face, of those of the cranium, strictly speaking, and of the hyoids. The second is composed of the bones of the cerebellum, of those of the palate, and of the parts of the larynx. The third, of the interparietals, the parietals, and the operculars."

But here Geoffroy Saint-Hilaire finds himself back with the logical consequence of the first volume of his *Anatomical Philosophy*, where he develops a comparison among vertebrates based on their anterior-posterior organization.

In spite of everything, we can pardon Cuvier's annoyance at intuitions that, for him, are hazardous and ill-founded. In a note in his second memoir, Geoffroy Saint-Hilaire reports Cuvier's indignation:

> Your memoir on the skeleton of the insects lacks logic from beginning to end, said M. Cuvier, expressing himself on my first work, with an extreme vivacity, before the professors of the Museum. You compare things that are in no way susceptible of comparison. There is nothing in common, absolutely nothing, between the insects and the vertebrates: at most one point, animality.

This was indeed an annihilation of Geoffroy Saint-Hilaire's undertaking.

In his third memoir, Geoffroy Saint-Hilaire even goes so far as to compare the ribs of vertebrates with the legs of insects and crustaceans. His work is particularly serious: he asks Michel Eugène Chevreul (1786–1889) and Jöns Jakob Berzelius (1779–1848) to make a comparative chemical analysis of the bones of vertebrates and the carapace of certain crustaceans, like the lobster and the edible crab. Establishing the presence of calcium phosphate, magnesium phosphate, and calcium carbonate in the two types of organs, he deduces an identity of nature: "Speaking *generically*, these are bones, or parts analogous to one another as bony substances." At the time, this result was of no interest. Still, through the quantitative analysis requested of the two eminent chemists,

Geoffroy Saint-Hilaire was anticipating what would later become comparative biochemistry.

The whole of Geoffroy Saint-Hilaire's elaborate work would not have been attacked if he had taken care not to call the insects' cuticle a vertebral column. But that is just what he wanted to do! For him, everything was only a pretext to unify the animal world. In fact, if the insects—and by extension the arthropods—were found to be vertebrates, Geoffroy Saint-Hilaire would then see only two great divisions in the animal world. Cuvier's four embranchements (vertebrates, mollusks, articulates, and radiates) would then be arranged in the following manner:

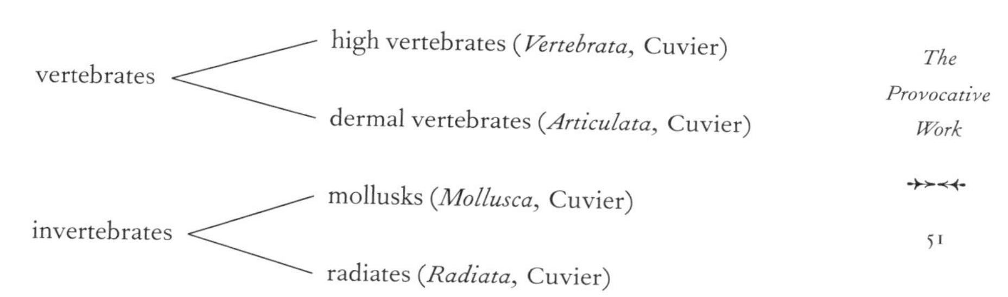

But finding points in common between insects and vertebrates following the anterior-posterior axis is not enough to unify their plans of organization completely. In fact, if we call the ventral face the one nearest the ground, the vertebrates—fishes and quadrupeds—have their nervous system in the dorsal position. Indeed, that is why they are called epineural. Arthropods, on the other hand, are hyponeural; that is to say, their central nervous system is in the ventral position. That will be the subject of the memoir of 1822, in which Geoffroy Saint-Hilaire postulates that the crustaceans—arthropods as well as insects—are reversed dorsoventrally with respect to the vertebrates. Geoffroy Saint-Hilaire resolves the problem in a purely geometrical manner. He proposes quite simply that the latter have "their backs below and their abdomens above." So would the vertebrates have the "good" orientation? We see him hesitate. His arguments are far from uninteresting. He bases them in turn on the biology of the vertebrates and of the arthropods. Thus, he proceeds to analyze the pyramidal decussation at the level of the medulla oblongata in the vertebrates as the result of a torsion of the body with respect to the head at the level of the neck. According to this vision, ancestrally, the abdomen of vertebrates would be their back, or the top of the head ancestrally in a ventral position! With the insects, he relies, for example, on the mode of locomotion of certain aquatic animals like the water scorpion or the water boatman, who swim, as it were, "wrong side up," that is, for Geoffroy Saint-Hilaire, following the position characteristic of vertebrates. Later, he would use the 1829 studies of the German embryologist Martin Heinrich Rathke (1793–1860),[3] published after his hypothesis, which show that in two arthropods, a spider and a crayfish, the reserves of the egg adhere to the region called dorsal by entomologists, while, in vertebrates, they are always ventral.

As the controversy develops, the comparison of vertebrates and insects sketched by Geoffroy Saint-Hilaire constitutes a decisive stage. In fact, if he is right, it is the whole system of Cuvier that breaks down, since it is Cuvier who insists that the four great embranchements are characterized by wholly irreconcilable plans of organization. Indeed, toward the end of his life, in 1838, Geoffroy Saint-Hilaire recognized that this was a key episode in the evolution of his thought:

> The time when I left the naturalists' common ground, I mean to say, their usual mode of description, of nomenclature and of classification, was chiefly in 1820, when I inquired whether and how there was a philosophical similarity of organization between the insects and the animals of the higher classes.[4]

What follows here is the complete text of the first *Memoir on the Organization of Insects*, the most important of the three, the one that proposes a comparison between the anterior-posterior organizations of insects and vertebrates. Then follows the article of 1822, *Considérations générales sur la vertèbre (General Considerations on the Vertebra)*, in which Geoffroy Saint-Hilaire restates his ideas and discusses the comparative dorsoventral orientation of vertebrates and arthropods.[5] Despite its length, I have included it without abridgement. It seems to me important to republish a text that is now referred to in the community of developmental biologists, and which is available at the moment only in very rare specialized libraries.[6] The nonprofessional reader is advised to go straight to the paragraph "De la vertèbre chez les insectes" ("On the Vertebra in Insects").

MEMOIRS
ON THE ORGANIZATION OF INSECTS

FIRST MEMOIR

On a skeleton in the insects, in which all the pieces, identical among one another in all the different orders of the entomological system, correspond to each of the bones of the skeleton in the higher classes.

Read at the Academy of Sciences, January 3, 1820

BY M. GEOFFROY SAINT-HILAIRE

My learned colleague M. Latreille, wishing to gain an idea of the whole range, as tools of investigation, of the two principles whose support, under the name of *theory of analogues* and *law of connections*, I had gained for my new doctrine, had undertaken some time ago to make an application of these principles to the insects. A circumstance that I regard as a most agreeable recompense for my work has come to give to that first plan the character of a consequence. That is the distribution made, a month ago, to all the members of the Academy of a brochure containing an analysis of my work. I permit myself to recall this little event in order to have and to seize the occasion of acknowledging all my gratitude to the author, M. Flourens, doctor of medicine, a young physiologist of a cast of mind to make one predict that from his first appearance he will rank among the first masters of science. It is in drawing his inspiration from that writing—it is, I say, in these circumstances that M. Latreille occupied himself a week ago with reducing to the same law of conformation the locomotive organs of the true insects, the arachnids, and the crustaceans.*

* The true insects have four wings and six legs, and the crustaceans ten legs in all. Do the wings of insects come from the development of certain tracheae, or do the ten legs of crustaceans rather correspond to the ten members of insects, legs and wings? Such are

Fundamental ideas are so called, not only because of their greatness, but because they produce a kind of alertness and hence are productive. The research of my distinguished friend M. Latreille has thus in its turn caused me to think deeply. Would it be unreasonable to suppose the connections he has investigated from insect to insect observable in insects in relation to the vertebrates? How many names there are, taken from the higher classes, and which today form part of the entomological dictionary—like mouth, head, eyes, thorax, abdomen, thigh, hip, tibia, etc.! To recall this alone: how many questions are implicit in this statement? Indeed, what reasons would have necessitated recurrence to these old denominations? Was it inspiration that led to it, or instead was the usage worked out rationally, through an exact determination of each part? Attend to this well: it would not be possible, without implying a contradiction, to tax with rashness, or with overambitious intentions, the project of comparing with one another the insects and the animals of the higher classes, under the pretext that those beings are much too far apart. For then, for the same reason, it would be necessary to go back and consider them from a general point of view. In fact, it would be necessary to have recourse as soon as possible to that remedy, as the only way of reducing to their correct value, through knowledge of their cause, the alleged relation that could not fail to be assumed between things called by the same name.

I had no sooner made these reflections than I found myself engaged in this investigation, without being able to remain the master of my subject. I was carried away by my two principles: anticipation of the existence of the same materials in all these animals, anticipation of the order of their arrangement, of that of their relations, and of their mutual dependence. In fact, I had never stopped proceeding under that great conception of nature: *the unity of organic composition*, and, in the confidence given me by that fundamental truth, I had to embrace my problem in its highest generality, advancing on each part, I do not say without hesitation, but nevertheless advancing there firmly, and without all the marvels and dissimulations that forms and functions, in their infinite variety, would offer me at every step.

It is to be expected that in this first work I would have only principal results to present. I reserve the demonstrations for a sequence of memoirs in which I count on taking up each organ one after the other, and where I shall

---

the questions that M. Latreille discussed in a memoir read, the 27th of last December, to the Academy of Science, and which he answered by adopting the second of these hypotheses.

undertake to compare it, at first from insect to insect, and, second, for what it will offer me that corresponds in the various groups of oviparous vertebrates.

1. General Relations of Insects with One Another

It is truly remarkable that these chief relations in insects are given me by the relation of their bony parts, just as invariably as I have observed it in the vertebrates; and in fact you may read in my *Anatomical Philosophy*, p. 8, that every part of the skeleton possesses for itself an appanage of soft parts, muscles, nerves, and vessels; that the bones are like hollow sheaths or are arranged like a sort of ninepin. Formerly I had settled on this proposition only with regret. Although I had had to admit that reason told me to guard myself against it, I had to return to it, since I was constantly led to it by observation. It is a fact on which my new studies shed new light. As that would be to anticipate what is due to follow, I shall not provide the explanation at present. I ask only that attention be paid to the fact in itself, and to its practical consequences, in order to arrive at a just appreciation of the real relations of things.

Now, if I go to break up an insect, or, more exactly, to separate the parts at the natural joints, even without taking great care, I arrive at six segments, to wit:

A *first*, or what has until now been taken for the head.

A *second*, called corselet, but not always known by that name in all the orders, and for which, to prevent all new uncertainty, and so that we may not be imposed on by all its metamorphoses, I note the connection with the first pair of legs, which this segment always carries.

A *third*, to which I assign as chief character a base for the insertion of the first wings. When of very slight extent, this base has been called the *scutellum*. That is how it is in the coeloptera. When much enlarged, it exceeds the anterior piece in volume, and takes its name, that of corselet. The hemiptera are in this condition, chiefly the cicadas.

A *fourth*, or what is properly called the thorax, which most commonly carries the four posterior legs and the second pair of wings. This last is a circumstance of which I believe it is of great interest to assure oneself, following the observation that the two pairs of wings, so different in the coeloptera, and of such a difference, in fact, that the first of the two has received its own name, that of elytron, end up in the hymenoptera, or even more so in the lepidoptera, being completely similar in their forms, their uses, and even their varied colors.

A *fifth*, composed of the abdomen.

A *sixth* and last, which is composed of the final segment, and which usually carries several appendages.

Having thus divided the insect into six parts or principal regions, I do not, however, mean to say that each segment is not further divisible. On the contrary, the thorax is divided into two parts, into which, respectively, the two posterior pairs of legs are inserted. And the abdomen, for example, is usually divided into eight parts, very distinct in the orthoptera, the mantises, the [spectres??], etc.

What I understand by primary divisions is an order of association that makes a grouping of several parts into an organ, or, as the common expression goes, a case with compartments. Thus in the higher classes the vertebrae of the neck, those of the thorax, the lumbar vertebrae, the elementary bones of the hyoid and of the sternum, form a whole, a collection of parts devoted to one and the same end.

One might be tempted to stop here for a moment, and, by choosing the insects with the most monstrous appearance, the most striking at least, by the singularity of their anomalies, to prove that they can all be reduced to this common law of conformation. But I do not want to make my case here through the details. Besides, I am putting off until later an explanation of how, in a given family, the second segment happens to get itself extended along the whole animal, while elsewhere it is the third segment that is thus elongated. I shall limit myself here to warning that all the parts of so disproportionate a magnitude that they conceal all the others observe in their divergences an invariable order, and that they always remain faithful to the principle of connections by means of their bases, whose insertion at a determinate point never changes.

Today I can only present this whole marvelous order of facts. I shall explain them later and furnish a complete demonstration for each family.

2. *General Relations of Insects to Vertebrates*

When I pass in review the different orders of insects, I find this singular result: that each can be studied under the diverse organization of the various groups included under the name of oviparous vertebrates. I have indicated in my work how the numerous divisions of the class of reptiles are related, some more essentially to mammals, others to birds, yet others to fishes. "Strangers to one another," I said, "the reptiles nevertheless converge on a common center, not in any way because it attracts them, but because it does not repel them: they fall under the same considerations because of an impotence common to all of them, that of the organs of respiration."

In the same way, the different orders of insects seem to me to come from the different systems that characterize* the higher classes. The insects likewise converge on a single center, and are grouped together because of a similar impotence in all of them, resulting from the lack of an agent to determine the arterial circulation. From the absence of a heart or its equivalent as left ventricle, which can send to a distant point a nutritive fluid composed of extremely mobile molecules—from this fundamental circumstance, it follows that there is only one mechanism, instead of several as in the vertebrates, only one mechanism for the distribution of the formative elements of the organs. The point of departure for this distribution consists in all the nerve ganglia of the spinal extension; and as the first acts in the development of the embryo concern this extension, the insects have a beginning that resembles that of the vertebrates even before there appear the slightest traces of a vertebral column.

The difference of the former from the latter is that, for lack of a centralized muscular force, of a second power for a second circulation, they reach their full existence under the same influence under which they began. The nervous system is the unique generator of organic materials, when, in the higher classes, we see it perform this service for the arterial system. The nervous system diffuses its materials; and spreads them out all around its axis, so that, step by step, the progressive development of whatever constitutes the organs of insects continues to take place within the spinal column. If this is granted, all the anomalies are explained, all the unknowns of this peculiar problem are unveiled: we find in the insects, contained all together in the same tube, not only their spinal cord, but all their abdominal organs.

If that is the case, we will no longer be surprised at the fact that in this other system of organization we find the spinal extension in an inferior position, and that the skeleton is cast out, so to speak, and actually becomes the integument, the superficial coverings of these animals. The tortoise already has its whole trunk enclosed in what constitutes its bony part. Its vertebral column is altered and formed of vertebrae, which are joined only at some points. If this anomaly were carried even further, the vertebrae would be too thin; they would open and let the spinal extension drop into the abdominal area.

From these facts it follows that the insects are *vertebrates*. And if every-

* I have drawn the information that I am at this moment applying to the insects from the unpublished work of M. Dr. Serres. His *Lois de l'ostéogénie* (*Laws of Osteogenesis*), which that learned anatomist will not delay in making public, in fact contain very new and very profound views on the formation of the fetus.

thing can be reduced to a vertebra, it is in the insects that this proposition is most plainly evident. In the last analysis, we arrive at this result: every animal lives inside or outside its vertebral column. We have in fact this great characteristic for distinguishing from now on the old from the new vertebrates, who will take their place immediately after the former ones.

But for this proposition to be rigorously deduced, we must be able, once the key has been found, to reduce to the itemized forms of the higher vertebrates all the parts of which the insects are composed. The thing is easy, and, I may add, it is at present perceived and basically decided.

To keep this memoir as general as I had intended, I shall only indicate my principal results and give a kind of recapitulation of the subjects that I propose to treat in what follows.

The *first* segment of the body of the insect does not correspond to the whole of the head of the vertebrate, but is composed of the bones of the face, of those of the cranium, strictly speaking, and of the hyoids.

The *second* is formed by the bones of the cerebellum, those of the palate and parts of the larynx.

The *third*, of the interparietals, the parietals and the operculars.

Thus, the three anterior segments stem from a dismemberment of the skull of vertebrates. MM. Oken,* Spix,† Meckel,‡ and earlier, but much more vaguely, Kielmeyer, J. P. Frank,§ and first of all Burdin** have noticed a striking analogy between some parts of the skull and the vertebrae. Bringing more vivacity and rigor to this idea, M. Blainville†† has also announced, on this subject, that he could demonstrate that the head of vertebrates is composed of a series of fused vertebrae, each developed in proportion to the nervous system that it encloses. The preceding remarks, from which it follows that this material dismemberment is visible in the insects, will furnish proof for this insight, and will change into a scientific fact the ingenious ideas of these celebrated German and French physiologists.

The *three segments that follow* correspond to our trunk and, like it, are divisible into the *thorax*, *abdomen*, and *coccyx*. It is only the posterior wings that

* *Ueber die Bedeutung, etc., ein* Programm (Jena: Quarto, 1807).

† *Cephalogenesis, etc.* (Munich, 1815).

‡ *Beyträge zur vergleichenden Anatomie*, vol. 1, p. 34.

§ *Epist. de cur. morb.*, vol. 2, p. 42.

** Cours d'études médicales, vol. 1, p. 16.

†† Bulletin des Sciences, by the Soc. Phil., July 1816, p. 108.

are difficult to reduce; we shall prove that they are analogous* to the swim bladder of the fishes or, what comes to the same thing, to the air sacs of birds.† At the moment of the transformation of the pupa into a perfect insect, the wing is a sort of bladder, a pouch with appreciable empty space;‡ when they lose their moisture their membranes collapse, come together, and extend so as to form a single plate.

Finally, we find among the insects an auditory or branchial opening and, on the flanks of their abdomen, a series of perforations, called stigmata, entirely analogous to the openings spread along the length of the lateral line of fishes. These perforations are so many outlets to carry to the outside the secretion of a long glandular apparatus existing below the lateral line. In the same way, birds also have, though in interrupted sections, a great glandular apparatus from the tongue to the anus; and with regard to mammals, I have shown what remains of it in the hypochondria of shrew mice.§

I have been able, in this first work, only to glance much too rapidly at all the organs and their correspondences. It is with the demonstration of all these pronouncements that I have promised to concern myself in the subsequent memoirs. I shall therefore confine myself today to announcing that in detail

*For my part, and by a different path, I had arrived at the following results, published eight months ago by M. de Blainville:

> The respiratory apparatus in the oviparous vertebrates is composed of two parts, which are distinct up to a certain point; the anterior one, always vascular, and the other, posterior part, often vesicular. Thus, the gills of fishes, the lungs of birds, the anterior part of the lungs of snakes, belong to the first, and the swim bladder of fishes, the large lateral air sacs of birds, the posterior part of the lungs of snakes, the lungs of olms form the second." (*Journal de physique*, May 1819, p. 400)

†In the memoir that we referred to above, M. Latreille had conceived with respect to these wings the happy idea that they could be derived from the tracheal tissue: "The wings of insects," he said in that memoir, "would they be the tracheal appendages which, in virtue of their robust muscles, of the fineness of the substance of which they were formed, of their aerial veins of the extent of their surfaces, would possess that property that their designation indicates?"

‡*Mémoires du Muséum d'Histoire Naturelle*, vol. 1, p. 301.

§I have, preserved in liquid, a coleopteran, of the species nasicorn beetle, that was placed there at the moment when it was shedding, and some of whose wings, shaped like a pouch, had received in its interior and retains, part of the liquid in which the animal had been immersed.

each part of the insects finds its like in the vertebrates, that it is always there in its place, and that it there remains faithful to at least one of its functions.

Another consideration to offer is this: that we arrive, doubtless against all expectations, at the demonstration of this fact: to wit, that the lungs, the heart, and the whole arterial apparatus proceed to efface themselves more and more starting with the higher animals down to the insects, when, on the contrary, the skeleton persists to the point where finally, in those same insects, it shows itself in an integral condition that truly gives much food for thought.

It was formerly a pretty widespread opinion that the skin, thick and as it were ossified in the insects, became for them a kind of skeleton: the muscles took their points of attachment above it; and M. de Blainville (preceded, however, by Willis, Lyonnet, and de Geer*)—M. de Blainville, while remaining attached to that same idea of an ossified skin, nevertheless advanced to a harmony of relations more real and more profound when he proposed to divide animals according to the consideration of whether they are articulated inside or outside.†

I shall conclude with one last observation.§ Analogy is advisory: let us reflect on this well, for fear of its consequences or its abuse. It is for having yielded too easily to its seductive fascination that, in the highest parts of science, we seem to have emitted opinions of the greatest diversity. Indeed, what a succession of different systems on the course of the blood, for example,

*Insects have a tough and flexible skin: hard, scaly, and as it were crustaceous in some species. It is to its interior surface that the muscles of locomotion are attached. Thus the insects would have, so to speak, their bones on the exterior, while other animals would have them *inside* their body. (De Geer, *Mémoires pour servir à l'histoire naturelle des insectes,* 1771, vol. 2, p. 2). Still earlier, and in more precise terms, Willis, in speaking of the crayfish, had already said: *Quoad membra et partes motrices, non ossa teguntur carnibus, sed carnes ossibus* (As to the members and mobile parts, the bones are not covered by the flesh, but the flesh by the bones). (*De anima brutorum,* 1692, p. 11).

†All the animals of my first type, says M. de Blainville, are either articulated on the interior, the *vertebrates,* or articulated on the exterior, the *invertebrates.* My principles for a new distribution lead me to draw these characters only from the organs of locomotion, or rather, from the combination of the *different appendages* by which each segment of the body can be accompanied (*Prodrome d'une nouvelle distribution des êtres. Bulletin de Sciences, par la Societé philomatique,* July 1816, pp. 107–8, 123). Such, until this moment, was M. de Blainville's whole doctrine on the insects; one can rest assured that after he reads this memoir he will know and say more.

§This peroration, written several days after the reading of the memoir, was communicated to the Academy at its meeting of January 10.

on the changes in respiration and the phenomena related to it! What is there more curious than the literary history of physiology envisaged in this manner? Although I was wholly devoted to this enterprise, it is nevertheless only in subjecting myself to the most serious misgivings about the future of entomological research that I am publishing today its first and principal results. The age that follows me and that will soon overtake me (for, in these times of such great activity for the works of the mind, the periods of science come close together and crowd on one another in virtue of the competition of a greater number of initiates to its mysteries)—that age will be able to attest to new facts and new experiments so as to see and to feel differently from the way I do today.

It is clear that I am not reporting these observations so that they may profit persons who are of a mature age: a person who has received the lessons of a long experience is immune from all seduction. I am addressing myself to youth, naturally avid for novelties. My honesty in the sciences, my love for the truth, and the misgivings that I have just refused to dissimulate commit me to warning this interesting youth against my own results. I can give them no greater sign of regard than by telling them why they should become impassioned for views they would be disposed to consider of great philosophical interest. The reason is an absolute condemnation of these same views* pronounced (with a

*Yet why should I not report here that that celebrated naturalist was not always so remote from this system of philosophy? If it was necessary to seek a recommendation for my works beyond their intrinsic value, could I desire one more powerful than the following passage, inspired twelve years ago in that savant by my own essays:

> M. Geoffroy has presented to the class fragments of a great work that he has undertaken on comparative osteology, in which he seeks to carry further than has been done until now the analogies among the corresponding parts of the various vertebrates, analogies that Aristotle has already recognized, and on which he had founded his admirable works of natural history, but which had not yet been followed as far as they deserved, despite the great number of works of which they had been the object. In fact, these pieces, these parts of organs, which recur as more or less similar in number, in position, despite all variations of size and usage, and against all apparent final causes, must necessarily depend on efficient and formative causes: they must belong to the most primitive means that nature employs. And if we can flatter ourselves that we can ever shed some light on the origin of organized bodies, that most obscure, more mysterious point of all natural history, it is, as it seems to us, from these analogies of structure that the first sparks may fly. . . . It is in bringing artfully together species often far apart, it is in trying to seize on some fixed points in

little violence, doubtless too much) by the chief of the modern school, by the greatest of the naturalists of our age.

If I stopped there, I would perhaps be suspected of having wished, for delicate minds, to have placed in that phrase an irony that would be no less outrageous for being presented in too flattering a guise. For indeed, it would be thought that, if a philosophical doubt sometimes makes me retreat from my own judgments, I have nevertheless conceived the idea that my aphorisms resemble the facts in the most appropriate forms in the present circumstances; otherwise I would not have written this memoir.

No, I have not wanted to wound an old friend. I have remained the same with respect to him, always equally devoted, as at the time when everything was on good terms between us and we shared everything. What M. Cuvier is condemning at this moment is the totality of my views, that is, my whole philosophy, which he has already made the subject of his criticisms* in the analyses of the academic sessions of 1817 and 1818. But what does this divergence of opinions really prove? Only that M. Cuvier and I think differently about theories: in that we do no more than reproduce respectively the two forms under which the human mind has always proceeded. The nuances† of these two ways of seeing and perceiving the facts appear everywhere where the judgment of men intervenes.

Finally, what had fixed me in these reflections was that, when I entered a group (at the home of Dr. Portal) two days after I had communicated to the Academy the first part of this essay, I was greeted with the words: "A young physician has just left here, whom the reading of your memoir has thrown into a delirium of enthusiasm."

I judged, at that moment, that it would be necessary to remind young

---

that crowd of apparent variations of beings, it is in pursuing each organ with constancy in all its displacements, that M. Geoffroy has come to establish new analogies. (Cuvier, *Analyses des travaux de l'Institut* [1807], 7)

*I seem to have noticed his criticisms only today; at least, I am speaking of them for the first time. As for me, I think that if I had the honor to be secretary of an Academy, I would state the facts, without distributing praise or blame. Without making judgments himself, the speaker of a learned society could in fact, limit himself to the exposition of the necessary documents; and for the rest, it would be well if he would submit his judgment to the public, the master of us all.

†These nuances hold chiefly for our point of departure, for the divergence of our ideas in the too-famous question of the preexistence of germs.

people that there was some wisdom in defending oneself against first impressions: they are durable only when they are thought through. A judge demands to see the parts of a trial, he examines them, and decides.

Let the new views set forth in this memoir be examined from two sides. I wish nothing more ardently than to see an enlightened discussion of them emerging.

GENERAL CONSIDERATIONS ON

THE VERTEBRA

BY M. GEOFFROY SAINT-HILAIRE

I have been asked to accompany this memoir with a plate.

In concerning myself with disposing of the parts, I have not limited myself only to the facts concerning this memoir, *the spinous processes of the dorsal vertebrae;* I have allowed myself to go further, to the idea of giving all the parts of a fish vertebra. Besides finding the advantage of presenting more clearly the spirit of my nomenclature, I had another motive as well.

Several scholars from Germany and England have done me the honor of attending the lectures that I gave in 1820 at the Faculty of Sciences, which had as their subject new considerations on the vertebra. Thus my work on this subject is already known to anatomists. I did not want it to happen that, when I had meditated sufficiently on the ideas I was developing and was thinking myself ready to publish them, it could be objected to me that something pretty similar already exists in foreign reports. In short, I must, and I wish to, avoid having what happened once occur a second time.

And in fact, in another course, in 1809, at the Royal Garden, I had given all my doctrine, not yet published, on the comparative osteology of the vertebrates. Everyone knows that it is permissible for hearers at the lectures given in the Museum to sign in on a register kept for that purpose. The inscriptions in my register for 1809 open with a name that I hold it an honor to see included there, that of M. Jean Spix. Six years later that learned naturalist published in Munich a large and important work entitled *Cephalogenesis,* etc. It was at this period that I finally thought my ideas on general osteology had acquired enough maturity to be presented to the public. I collected them in the first volume of my *Anatomical Philosophy,* which I published at the beginning of 1818. Some of the analogies that I had found and which are reported in that volume were then thought to have a striking conformity with those of *Cephalogenesis.*

Since there was very lively insistence on these remarks in Germany, I was obliged, in order to remove a suspicion that would have wounded me, to establish that at the Royal Garden we knew of the existence of the *Cephalogenesis* only at the end of 1818, when M. Cuvier returned from a trip to England. My observations concerning this appeared in *Isis,* a scientific review published in Leipzig.

Such are my motives for giving at this time my generalizations on the vertebra and for deviating from a usage wisely established in scientific method: a method that demands first the exposition of particular facts before the admission of the relations that apply to them and which constitute general facts. Since I was occupied with finishing my work on monsters, and had decided to follow it with another, begun a long time ago, on the sexual organs, I could not predict the time when I would take up again my work on the vertebra. Thus what I am about to present today is intended to fix a date and to assure my rights to the ownership of considerations that I believe to be important.

To procure for my considerations a basis from which I could take my point of departure, I concerned myself with finding a vertebra that would be kept within average proportions, which, since it was situated at equal distances in the graduated series of developments, would retain traces of the first formations and would at the same time express some indications of the subsequent stages, and which, finally, had reproduced its different parts in homogeneous forms.

I stopped at a bony section of a young plaice as offering me the conjunction of all these circumstances. It is this early age that is in question in our markets under the name of dab, and under that of *Pleuronectes rhombeus* in the works of naturalists. All the materials of a vertebra are to be seen very clearly in this section; at the same time they are presented under a form with perfect regularity. So much so that we may fear there is nothing like it in living nature. The nucleus of this vertebra is so exactly in the center of the system, and every part leaves it by rays that correspond in such a marvelous manner, that we are tempted to see in the drawing that represents these arrangements only the sketch of an ideal type. We shall see below how in fact a fish that swims on its side, and whose tail extends horizontally on the surface of the water, furnishes that intermediate example that we had set out to find. A pleuronect does really belong to the animals of the first series (it is one of its last segments), and by the anomalies of its way of swimming it opens the series of inferior classes, almost infinite in number and in diversified forms.

Concerning the Vertebral Nucleus

Kerkring, and all the anatomists concerned with osteogenesis, have seen the vertebral body as wholly isolated, in the human fetus of eight or nine months, and hence separated well before that period. I show it detached in a bovine fetus at an age proportionately less advanced, plate V, figure 8, and in the plaice, figure 2. I was the first* to present the vertebral body as primitively tubular. It remains in that state for as long as the fluids that furnish nutriment to the animal that is developing are mixed together—a fact that has its analogue in the albuminous liquid of a bird's egg at the beginning of incubation. The course of development leads to the ichthyological combination, in which the vertebral tube, filled up with concentric layers, is no longer finally or for so long a time or even always perforated only at its center. For the disposition of these facts, see plate V, figures 1, 2, and 4. My research on the spinal column of the lamprey and the examination I have made of the fluids contained in the vertebral nuclei have fixed my attention on these important considerations and above all have made me appreciate their physiological value.

For the vertebral nucleus, this is an intermediate state between the tubular disposition of the first formations and the arrangement presented by more advanced fetuses or by young fishes: a membranous nucleus while it is only a tube, it soon gives way and folds up through the successive diminution and retreat of the fluids that were distending it. This takes account of the lateral and spiny projections that the nucleus exhibits, figure 1. Confined as by a belt at the middle at the same time that it is fixed at its extremities, this tube takes the form of the water clocks or hourglasses in use aboard ships and looks like the product of two cones united at their apex. This form is rendered with sufficient accuracy in figure 3 and figure 6. It is then that the interior of the vertebral body is still further obliterated, until it is finally closed completely. In this last state it is no longer more than a completely filled tuberosity, or that same bony disk with which they are concerned in human anatomy only to recall its shape, that of the section of a column.

*See my *Third Memoir on the Organization of Insects*, in the *Journal complémentaire*, etc., vol. 6, p. 146. See also the article by the late M. Presle-Duplessis, included in the *General Annals of Brussels*, vol. 8, p. 374, in which that interesting young man takes account of my later work, which had as its subject the vertebra in the lampreys.

Concerning the Lateral Branches

When the nutritive liquid comes to abandon the central pouch that contained it,* or the center of the vertebral nucleus, it splits into two different systems and moves into two other pouches that are situated along the length of the vertebral body, that is to say: one above and the other below. The neural arch forms the upper pouch and the hemal arch forms the lower pouch.

The presence in these locations of neural and vascular systems explains the permanence of the two cavities located in the vertical axis of the vertebra. For it is a similar cause that produces both of them; since it is to protect the circulation of certain fluids that other compartments are needed than that of the intervertebral canal. Similar means provide for this; never mind the different quality of the fluids that are poured in there. Thus, where there is the intervention of two pouches, these pouches are protected and separated by two bony cavities, for the formation of which similar elements concur above and below.

In fact it is the perials, *e e*,† that cap the neural system above, as it is two exactly similar pieces, *o o*, that take hold of the vascular system below and in their way contribute to closing it off. In mammals, where the spinal marrow is of a certain volume, the two upper ossicles, *e e*, or the perials, extend for their whole length around the neural process. It is entirely different in the fishes, in the case where our considerations apply to one of the vertebrae of the abdominal region. In this place the spinal cord is reduced to the state of a slender thread; it is no longer the perials in their whole length, but only a small part of them that encloses it.

But as one dimension never loses anything without giving place to the augmentation of the opposite dimension, so the perials of fishes, instead of being short and thick, as in the example of figure 7, are slender, but then disproportionately elongated, pressed one on the other, except for the points of their surface, which are in contact with the spinal marrow. They are joined to-

*This is not the place to explain how it happens that a system of nutritive vessels is obliterated and how it is replaced by other systems that are newly produced. I can only suggest it according to a particular well-known case. The whole dental apparatus, its vessels, its nerves, the teeth themselves, are destroyed and disappear at a certain stage, under the action and through the efforts of new nervous and vascular ramifications.

†The term "périal," translated here as "perial," like those that follow, appears to be Geoffroy's invention; it seemed best just to provide corresponding English terms. The perial in fact refers to the left and right sides of the neural arch. *Trans.*

gether promptly—a temporal arrangement that is a uniquely ichthyological circumstance. This holds, moreover, only for the postabdominal vertebrae.

For if it occurs, even in the fishes, that the spinal cord is enlarged, the perials, *e e*, no longer suffice to surround this new content and provide for it a cavity of a proportional capacity. What happens then is that first they move apart; but furthermore, that they become aligned and give assistance to two exterior pieces *a′* and *a″*, figures 1, 3, 4, 6, and 7. These four pieces together form three quarters of a circle, which is completed, as to the missing quarter, by the vertebral body. It is in this way that all the cranial vertebrae are disposed; for it is only when the cerebrospinal system is prolonged in the cranium that it acquires a larger volume and that its container, or the cerebral cavity, demands that it make use of all possible resources.

What we have just said about the upper parts applies in all respects to the branches of the lower region. In effect, the vascular system is no longer formed exclusively, as in the examples of figures 1 and 2, by the arterial trunk alone, but, on the contrary, when it comes to acquire many dependencies (such as are the organs of digestion or those of respiration), the paraals, *o o*,* behave as we have seen the perials doing: renouncing the union that had led to their fusion in the postabdominal vertebrae, these congeneric pieces move apart and become long appendages known as *vertebral ribs*. I remind my readers that I had already named these earlier, from the fact that these ribs are articulated to the body of the vertebra itself; I show these long ribs in place: see *o o*, figures 3 and 5.

If nothing has been added to the vascular system except the parts most immediately dependent on it, such, for example, as the pulmonary vessels, the bony appendages are prolonged far enough to meet, and in converging on the bones of the sternum, they describe three quarters of a circle, which, just as in the case of the cranial vertebrae, is completed by the vertebral body. Thus there is established underneath, with respect to the rib cage, what we have seen establishing itself above with respect to the cerebral cavity.

For the ribs, or to include them under a name that recalls their higher condition of generality, the paraals are no longer sufficient to produce so large a circle. Just as in the upper region, these pieces are followed and assisted by others that are fastened to them and which are our ossicles *u′* and *u″*, that is to say the two cataals,† which, since they have to do with the sternum, I had named *sternal ribs*.

* That is, the left and right sides of the hemal arch. *Trans.*

† That is, ventral (or pleural) rib and dorsal (or epipleural) rib (or hemal radials). *Trans.*

If, on the contrary, there is too much space to traverse in front of the abdominal organs, the paraals,* or the vertebral ribs, remain floating laterally; such are the pieces, *o o*, figures 3 and 5. The cataals or the sternal ribs exist nonetheless; but in this case they are in an undecided situation, spreading arbitrarily at different points of the vertebral rib, or placing themselves now at one end and now at another; the letters *u u* in the same figures 3 and 5 show two of these combinations.

Finally, the epials above† and the cataals below, finding themselves in the abdominal section, of no use for the partitioning of the neural and vascular systems (which we have seen is sufficiently provided for by the perials) and the paraals are deprived of their general function and in this way are bones without a preassigned use. Ready, then, to take on all sorts of service, they are repugnant to nothing and turn out to be susceptible of the most bizarre forms.

It is in these circumstances that, to serve as rods for the dorsal and anal fins, one of these pieces mounts on the other, one maintains itself inside, while the other launches itself outside, and together they compose a small apparatus, where it is no longer as congeners that they are interacting. Their functions change as their relations do: one has the movement of an arrow oscillating on a pivot, while the other is held still to support the other's efforts.

But whatever happens, it is always the same pieces, whether it is a case of congeneric bones, placed, one at the right and the other at the left, or of bones superimposed on one another, and thus distinct in their form as well as in their function. Under this last condition, these pieces, chiefly the exterior bone called a *ray*, have interested ichthyologists very much. Before this recent work, no one had ever imagined that they could find their analogues elsewhere, toward the crest of the vertebrae, where it is known that they do not form any protrusion to the outside; so that the names that serve to distinguish them indicate only one particular manifestation of them. I thought that by using a significative preposition placed before the words *epials* and *cataals*, I would simultaneously express on the one hand the origin and the destination common to these pieces, when they belong to an apparatus within which the most important phenomena of life are carried out, and on the other, their variation and their isolation in the case where one of the pieces is separated and distinguished

* That is, neural radials. *Trans.*

† That is, neural radials. *Trans.*

from its congener. That is the object of the following denominations: *pro-epial, in-epial, in-cataal,* and *pro-cataal.*

There are other flatfish, like the *Zeus romer, Centriscus scolopas,* and *Scaris siganus,* whose vertebral appendages form complete hoops around their abdominal organs. The paraals, or the vertebral ribs, are prolonged in such a way that they converge on the ventral ridge. The caraals still exist, but they are seen outside; and they remain short exterior spurs, whose origin it has not yet occurred to any one to investigate.

Of the Vertebra in Insects

One of the greatest joys I have experienced in my life was granted me through the happiness of my insights into the organization of insects. Once my mind had grasped this organization as being one of the first stages of the embryo, as offering among others that stage at which the principal vital organs are assembled in a unique center, I could say with assurance: "Insects live within their vertebral column, like mollusks in the interior of their shell; a true skeleton for the latter, a kind of contracted skeleton."* I could say, and I opened the year 1820 with that declaration, I could say that insects formed a further class of vertebrates, that consequently they were subject to the general law of the uniformity of organization, and as intermediates between the vertebrates of the higher classes and the most feebly endowed animals, the whole mystery of their affinity with the former as well as with the latter, with the mollusks and in general with all the beings of nature, was revealed. These propositions caused consternation; I have not been able to forget it. No transition had prepared minds for this: their novelty gave rise to very lively objections, within as well as outside the circle of naturalists. Contrary ideas had on their side the sanction of time, matters of linguistic usage and theories already organized. What would not be difficult in itself was to assert one's disapproval.† I kept my patience, already daring, however, to take the opposing side, and to make heard the fa-

*See my second memoir on the insects, vol. 6, p. 35 of the *Journal complémentaire.*

†I had just delivered at the Academy of Sciences a report on the entomological research of M. Audoiuin. This work was accorded only a conditional approval; I was obliged to withdraw the term *vertebra* as applied to a unit in the body of insects. The celebrated philosopher of Pisa, also constrained to disavow his discovery on the movement of the earth, at once went back on his word, by letting fall the phrase: *but nevertheless it does move.* Could I in my turn declare false what I believed to be true? How I regretted knowing too much at that moment, not being able to give the assent demanded of me, and thus

mous cry of nonconviction: *e pur si muove*. However, I had not waited too long before those very persons whose objections had been the liveliest undertook the role of teaching the same ideas, with some reservations, or rather with some changes in terminology.*

But stop; let us forget these debates; and in strength, with a deeper sense of the facts, let us establish once again, and by new proofs, now wholly decisive, that insects occupy a place in the series of the ages and of the developments of the higher vertebrates, that is to say, that they actualize one of the conditions of their embryo, as the fishes do for one of those of their fetal age.

What forms the outstanding character of the insects and what chiefly distinguishes them from the higher vertebrates is that the bony pouch of their first stage is not replaced by two specialized pouches, that it continually maintains itself, and that the vertebral tube of which it is composed, far from experiencing its ordinary metamorphosis, that is to say, of filling up and organizing itself in a nucleus, around which and on which all the organs are constructed and on which they depend, is on the contrary under the necessity of enlarging more and more under the action and through the vitality of the nutritive fluids, to the extent to which the organic systems of the interior themselves acquire stability

seeming to lack respect for so august an assembly! M. Duméril, who was charged with the commission withdrawn from me, was approved. Our two reports can be compared; I printed mine in the *Journal complémentaire*, vol. 6, p. 36.

*It is in the nature of things that a discovery, before being irrevocably acquired for its author, goes through two successive tests. People start by denying that the discovery is real; is it authenticated? Either through public action or by insinuation, it comes to be attributed to one of the ancients.

It must doubtless be admitted that my idea of *a vertebra in the insects* is now touching on its second period. For I read in volume 8 of our Memoirs, p. 469, that "the term vertebra in respect to insects is employed by Wotton, p. 175 of his book, *Differentiis Animalium*." In fact, that author does consider certain links in the body of insects as vertebrae, in the context of their nesting and of their mobility; that is in fact all that that passage means: *implicatis flexilibus vertebris* (with the flexible vertebrae folded in). Wotton could not attach to the term "vertebra" the same sense that I do, for some lines later he applies to the subject of insects the ideas of the Aristotelian school, which are still the reigning opinions: *osse carent*, he says, *osse carent exsanguinea omnia, sed neque spinam habent, ut pisces* (*All the bloodless animals lack bones, nor do they have a spine, as fishes do*). However, the report that Wotton presented does honor to his sagacity, especially if we consider the time at which he wrote it, which was 1552. Wotton formulated his report with all the more good fortune, since, limited to the role of a scholar, of a commentator on the books of antiquity, he lacked altogether the sense of naturalists concerning the analogy of organs.

from their central pouch, and that they pursue their development and acquire volume.

In this case, it is not the vertebral tube of fishes that is filled by concentric layers, and which remains scarcely pierced at its center, it is a ring whose diameter equals the very size of the animal; from this follow several results that merit attention.

The first, which follows entirely naturally from our law of the balancement of organs, is that the thickness of this ring or the solidity of the vertebral canal is in an inverse ratio to the length of its circumference.

The second, that the vertebral tube, finding itself at the limit of the derm, is immediately clothed by it.

The third, that this last circumstance, which the muscular powers contained in it cannot disturb, like all the other organs, in the vertebral nucleus, leads the two tubes, one inscribed in the other (the osseous and the epidermal) to unite and to become identical.

The fourth, that the respective volumes of the two tubes vary gradually, either in direct proportion or in inverse proportion, without the organization's experiencing a serious alteration or being perceptibly modified. Thus, in order that the dermic tissue should receive more nourishment than the osseous tissue, and acquire proportionately more thickness, we have the solid envelopes of the coeloptera, which a thoughtless attachment to ancient custom constantly and improperly calls *horny tissue;* or in order that, on the contrary, the osseous tissue should predominate over the epidermal, we have the more solid and more resistant shell of the crustaceans, that is to say, a true osseous system in the two contexts of organic structure and chemical composition.

A fifth and final result is that, all the nutritive fluids and the organs that they engender continuing to be concentrated in the vertical canal, no other canal is necessary outside it: consequently there are no more double pieces*

*It is a necessity in the insects that the vertebral branches be disposed along a single line. However, as if in this class, in which the number of species terrifies the imagination, it would be necessary (for that is really the idea that the forms of these animals, varied to infinity, give us) that all be combinations the mind could conceive be realized, we still meet here traces of the forms of organization proper to the higher animals. When it is, as for example in the crustacea,—when, I say, it is the odd and median piece to which the greatest quantity of the nutritive juices are diverted,—the lateral branches, receiving less nourishment, are no longer more than rudimentary parts, distinguished by entomologists under the name of false legs. Concentrated in themselves, these parts are positioned in relation to one another in the same way as in fishes. Beyond the cycleal or the odd piece there

that branch out below and above the vertebral body; no more compartments to enclose the neural system above and the vascular system below.

But in occupying ourselves with the higher vertebrates, we have noticed that the less space occupied by the partitions between the neural and the vascular systems, the fewer pieces it was provided with. We have in fact seen the order of degeneration that follows. If the spinal cord attains its *maximum* volume, as when it opens out in the cranium, the circle that surrounds it is composed above and on its sides of the four possible pieces (the two epials and the two perials, which are articulated to one another by their extremities. The same remarks and the same consequences hold with regard to the inferior bones of which the rib cage is formed. Does the spinal cord in the dorsal region, as in the steer (fig. 7), form a process that still receives enough nourishment? It needs only two pieces, *e e*, for its osseous girdle, pieces that at least here are applied in totality; and then the two other pieces, dismissed from this service, become the vertical ridge *a′* and *a″*, and thus take on other functions. Finally, it is yet another of the possible cases, of which we give an example in figures 1 and 2, that the spinal cord is attenuated and brought to the state of a slender cord; it is then too much for the two pieces *e e*, or the two perials, to embrace it: it suffices for one portion, a fifth part of the bones, to meet there, while the remaining four-fifths, which have become long threads buttressing one another, unite, fuse, and become one long process directed from the center to the circumference.

Continuing to descend the rungs of the ladder, we arrive at the insects, in which particular cavities to lodge the spinal marrow no longer exist. At the higher level, we have seen the paired pieces ceasing to appear as congeners and taking a place one above the other. Useless and foreign to a cavity that has become too small, they make their accessory functions a final object: they devote themselves entirely, for example, to furnishing a point of support for the muscles of locomotion, to favor overall movements, and to regulate equilibrium when at rest. Where there is no longer a cavity devoted to the neural system, it

---

are two pieces squatting at each side. Thus, these bones are arranged as if they had to serve to partition off, on the one side the spinal marrow, and on the other the great artery. I have represented this arrangement in Plate VI, figure 1. So that this abdominal vertebra of the lobster would be more easily comparable to the vertebra of the pleuronect that we have already encountered, I have placed it in the same situation. In Plate VI I have separated the upper part of the derm, which I have allowed to subsist on the interior branch. Observation, assisted by the use of the same letters for all the corresponding parts, will say more than all the details I could imagine.

can no longer be only two pieces, but the four parts of the cerebral cavity or of the rib cage, which stop being paired two to two belonging together as congeners. And what I have just said about the neural system as applicable on the one side holds for all the relations of the vascular system and the other parts of the opposite side, since the two systems are equally contained in the vertebral tube.

Do we arrive at this question from on high? As long as, in the insects, the vertebral nucleus keeps for good the disposition that it assumes from the first in the embryo, the form of a tube, and as long as this tube grows larger through the pressure of the essential organs of life, all of which develop in its interior, it follows that the other parts of the vertebra are on the outside only unimportant dependencies and could be appropriated only for locomotion. Do we then see our subject in its relations with the organisms of the immediately superior segment in the series? These are the ossicles that are in line, four for the top and four for the bottom, which analogy already told us we ought to find. Finally, do we see this subject in itself and independently of the presentiments and the relations that it gives rise to and that it brings to mind? Observation shows us, on the exterior of each vertebral tube and of each segment, a double series of pieces. Do we seek its use? It leaps to the eye that these pieces are used for locomotion. The connections? It is just as manifest that they have relations only with one another and with the muscles of locomotion; that is to say, and as a final consequence, that all of them and every part are, in form and in use, coordinated in such a way as to reproduce the ichthyological type and chiefly the postabdominal trunk, as we have shown in figures 1 and 2.

However, I must stop here in the face of a consideration that will not fail to be presented to me as a serious objection. Actually, I am comparing inclusively parts extended vertically, the vertebral appendages of fishes, with parts elongated on the sides and situated horizontally, with pieces that the entomologists call, in the decapod crustaceans, *true legs* or *false legs*, depending on whether or not they form part of the means of locomotion. Is there really something here pointing to a law of which I myself have celebrated the universal application, the principle of connections? At bottom, there is nothing real in this objection. It is, I say, among the cases in which the exceptions, assessed at their true value, turn out to be in favor of the rule. Thus the irregular movements of the moon owe it to a learned explication that they have become one of the most incontrovertible proofs of the system of the world, while Newton always found in them the character of a troubling objection.

I have never hesitated on the principle of connections: it has constantly served me as a rule; yet here is a circumstance where it seems to falsify results that would appear true in many other respects. These contradictions are the occasion of surprise on my part. However, continuing to consider the principle of connections as being our chief rule, as forming a principle that it is necessarily obligatory to apply in all possible cases, I examine the question of whether it has been given a correct application in the present circumstance. Well then, I notice that the different position of the vertebral appendages, observed in the central region of the back and of the abdomen of fishes and, on the contrary, on the sides of crustaceans, might indicate only a difference of position with respect to the ground. The fins of the former rise vertically and the feet of the latter are extended horizontally.

But do animals always have to move so as to present the same surface to the ground? The pleuronects—and I could present a much larger number of animals who are in this situation—the pleuronects reply in the negative. What our law of connections demands absolutely is that all the organs, in the interior as well as the exterior of the animal, be in the same relations with regard to one another; but it is indifferent in itself that the cavity containing them lies on the ground by applying one or the other of its surfaces. What relations of organization between man and the digitate mammals, and nevertheless, what differences in their posture! It is the same with the pleuronects and other fishes. The pleuronects swim lying on one of their sides, whence it follows that some of their fins, which are elsewhere directed vertically, like the dorsal, anal, and caudal fins, are in their case extended horizontally.

As far as the relations we propose to discover are concerned, this condition of existence is of very great importance. Suppose, with your mind fixed on this consideration, you wish to examine a crustacean. What do we see there, what do we find with respect to its position? An animal the same as the pleuronect, an organism that extends in the same way, to the right and to the left, the instruments it has as its disposal for its movement.

However, I cannot rest content with this relation, if it rests either on a simple appearance or only on a certain analogy of function: the principle of connections must lead us to judge the value of this consideration, either alone, or at least as of primary importance. Now, here is what the law we have invoked demands: it is that, if we have suspected with good reason that the crustacean makes of one of its sides its dorsal face and of the other side its ventral face—that is to say, if, in the manner of the pleuronect, it makes lateral (carrying them to the right and the left) parts that are vertical (upper and lower) in other fishes—then its tail has to be horizontal like that of the pleuronects. But that necessary consequence is indeed how things are: nobody is ignorant of it: for who has not noticed the tail of a crayfish?

The name given to this organ indicates that it is without hesitation that the tail of a crustacean has been related to that of a fish; but what has not been perceived, or even suspected, is, on the one hand, the relation of position, a persistent relation, of the tail of the one and the other with respect to the appendages of the vertebra; and, on the other hand, the relation of the smallest constituent elements, bearing on a similarity yet more real at bottom than it is manifest from the outside. The structure of a crustacean's tail is, in fact, in its essentials, absolutely that of the tail of fishes. The collection of nerves called *"horse's tail"* comes, in the crustaceans as well as in the fishes, to issue in long epidermic threads; as they terminate, in the whole-hoofed forms, in the bristles of their long tail. It is also the same structure as far as the form is concerned; but to convince oneself thoroughly of this, that is, so that it becomes a fact manifest to the eye, there is a distinction to be made. It is not all that is called the tail of a crustacean that answers to the last vertebra of a fish, or to the tail strictly speaking. It is exclusively its middle section: only that part forms the last vertebra of the skeleton of a crustacean.

In order that this proposition should be well understood, I have represented in plate VI, figure 2, the four last segments of the tail of a lobster. Designating each of these segments by the capital letters D E F G, and their corresponding appendages by the italicized letters *d e f g*, I make manifest to the observer the relations and dependencies of all these parts. The three portions *fff*, considered until this day to be the wings of the tail in the lobster, are at bottom only the appendages of the penultimate segment F″: that is what the connections and the articulation of F with *fff* demonstratively establish. As to the extraordinary volume of these appendages, that holds particularly in the neighborhood of a rudimentary segment: for in the organization as a whole, what one organ loses in development is claimed by the neighboring part.

The last segment corresponds at all points to the last vertebra of fishes: the crayfish furnishes the example most favorable to this system. The rays of the tail, their distribution, their nodosities, nothing is forgotten for the coccygeal bones of these animals to be an exact repetition of one another.

But more than that, what I wished principally to establish is the relation of the position of the tail of the crustacean to that of the tail of fishes. Figures 1 and 2, plate VI, no longer leave, I believe, any doubt in that respect.

Thus the principle of connections, which for a moment we supposed to be wanting, is truly not so: for it would be to make a false application of it to demand that the tail should conserve from one family to another its same relation to the ground. All that could be demanded would be that the vertebral nuclei of one and the same spinal column should have all their appendages directed to the same side, and we noticed that this is manifestly the case; whether the appendages of the dorsal vertebrae, the postabdominal and the coccygeal find themselves extended horizontally, as in the pleuronects and the crustacea, or the appendages and the tail are presented in a vertical situation, as in the majority of fishes. It is indeed true that up to now all these branches have been named differently (legs in the former and fins in the latter), following what could be perceived of the situation of these parts and what was thought of their use. But these are denominations that are faulty and that have led us into error, which was inevitable in an age when only the individual facts were studied and where it could not be suspected that they were bound by mutual interrelations.

However, it will be said, how admit that crustaceans, if it is true that their conditions of existence do not differ too much from those of the higher vertebrates, really swim on their sides? Won't it be necessary, on that supposition, that the whole of their sense organs, whose spheres of activity necessarily expand in all animals, and in the crustaceans are in fact exercised at the surface of the earth, should have suffered a kind of torsion and have crossed along with the rest of the spinal column? All animals, each one for itself, are in a theater of exploitation; and they do not succeed in assuring their well-being unless, through the scope of their senses, they reach the limits of their field of habitation.

I shall not develop this reasoning further; I shall reply as I did earlier. The possibility of this torsion is already demonstrated by the reality of this fact, evident in the pleuronects: they can have their eyes on the same side only by the crossing of the optic nerves and of all the parts related to them. Thus forewarned by these considerations and engaged in investigating what could be the case with the crustaceans, I proceeded to another fact of organization, to which I attach (and this is saying a lot) a much greater importance than to the vertebra in the insects. I shall seek only a confirmatory proof of my first facts; and not only is it given me, but I have just found that all the soft organs, that is to say, the principal organs of life, are reproduced in the crustaceans, and consequently in the insects, in the same order, in the same relations, and with the same arrangement as their analogues in the higher vertebrates. A single circumstance provided me with the solution to such a great problem, that is, the choice of a point of view from which I could look freshly at my subject. Thus all that organization of the lower animals, in appearance so confused, is disentangled, as happens with a skein of thread that is snarled, if one has the good fortune of keeping the principal thread in one's hand; this whole structure surprises by its extreme simplicity; and by allowing me to perceive the arrangement of a plan that differs only by degrees in the progressive order of developments, it leads me to reproduce the fundamental proposition of my writings; to repeat once more that organization is one, that it is uniform, and that it awaits only favorable circumstances to raise itself, by additions of parts, from the simplicity of its first compositions to the complication of beings that are at the top of the ladder.

I wanted to know if the crossing of the pyramidal bundles in mammals and birds was not due to a movement of torsion of the head, in a half-turn of conversion of the parts of the organism that were advanced with respect to its spinal column. According to my ideas on the vertebra in insects, the whole of the body of these animals forms a long and wide case composed of vertebral tubes or mobile rings. A cavity of this kind, although it is the essential organs of life that it contains, can behave in various ways with respect to the mass of these organs: the principle of connections demands only that all be in well-established relations with one another; but it would not derogate in the least from that principle if the whole totality had oscillated within the case.

According to this presentiment, to observe was already to have discovered all the relations that I was seeking to know about. I placed the animal, not as it is positioned relative to the ground, but as it suited me to see it in order to compare it with animals of the first rank. After I had destroyed all the vertebral tubes or the bony cases of the lobster I was experimenting on, I disposed of the rest of them in such a way as to see inside the spinal marrow. What was my surprise, and, I add, with what admiration was I not seized, in perceiving an arrangement that placed before my eyes all the organic systems of the lobster in the order in which they are arranged in mammals? Thus at the sides of the spinal marrow, I saw each and every one of the dorsal muscles; underneath were the digestive system and the thoracic organs, still lower, the heart and the whole vascular system, and finally, still further down, forming the last layer, each and every one of the abdominal muscles.

I shall not elaborate this point further, since I must return to these new and curious considerations: a new memoir will soon follow this one.

However, in what relations do these neural and vascular systems find themselves with respect to the case that contains them? In an inverse state, relative to the idea that we form for ourselves of the words *back* and *abdomen* (dorsal and ventral). Look at a crayfish turned on its back, and the whole order that

I have just noted is that of its various systems, as it is also that of the same systems in the higher vertebrates.

Thus the purely gratuitous supposition of a great sympathetic nerve in the insects taking the place of a spinal column collapses; a bizarre determination, which, it was believed, explained the *inferior* position of their neural process. It would be worth as much to admit the existence of a tree wholly of terminal branches and without a generative trunk that attached it to the ground, or that of a thing with a conclusion but no beginning.

There are a host of other consequences that flow from these first insights; but they do not belong to these general views. I shall give them later.

Nevertheless, it is the case that we have here certain facts and evident relations. However, not everything has yet been said about them. Some difficulties remain to be surmounted; but this time they are not inherent in the subject. The spirit of rivalry is implacable; and, such is its consequence, new vexations are inevitable.

Will others be disposed to acknowledge these new facts, if they fear to give, at their own expense, too great a share of glory to the discoverer? On the contrary, what resources one can provide for oneself! Are we not to validate both the prestige of ancient designations and the universal consent given to received ideas, and that good opinion of oneself, that we so easily believe we have the right to share, even to impose on others? All that will make the old prejudices on the insects last some six months. But time, which acts so powerfully and always efficaciously beyond the interests of rivalry, will set all things in their place. It is not without pain that truth is obliged to leave the well; but once outside, she shines with such brilliance that she shows herself and pleases universally.

Explanation of the Plates

In writing my abstract on the vertebra, I only wanted to give a very simple explanation of the plate that I had had made, in order to explain the spinous processes of the dorsal vertebrae; but finding myself insensibly engaged in an exposition of general views, I have produced a second memoir. Further, this second work necessitates a third, for the exposition of the facts that I have just announced concerning the relation of the organs lodged in the vertebral canal in the insects. I believe these facts are very instructive; and to give them with all the clarity desirable, I have had them represented in two further plates, which, I hope, can appear at the same time as the plate relative to the spinous processes of the vertebrates. However, the explanations that follow will be related only to the subjects treated in the two published memoirs, one on p. 71, and the other, p. 89.

PLATE V. Figure 1 represents the postabdominal vertebra of a young plaice, or dab, *Pleuronectes rhombeus.* There are only the double pieces *e e* branching above the neural system, and the double pieces *a a* comprising the vascular system, which are soon fused, each to its congener. The letters *e e* and *o o* indicate that two primitive materials supported, one on the other, the right hand congener on that on the left, so as to form afterward only one single piece.

I have used five vowels in italics to constitute the sign of the following designations.

Five names are in fact enough, if, except for the uneven piece, the vertebral pieces are paired off two by two; here is the table of them:

a′ *left* epial		*a″* *right* epial
e′ *left* perial		*e″* *right* perial
	i cycleal	
o′ *left* paraal		*o″* *right* paraal
u′ *left* cataal		*u″* *right* cataal

If these pieces are arranged in a unique series, I add to these primitive names a word that expresses their relations with respect to one another, as in the following example:

a′ pro-epial		*a″* in-epial
e′ meta-perial		*e″* cyclo-perial
	i cycleal (always an odd piece)	
a′ cyclo-perial		*o″* meta-paraal
u′ in-cataal		*u″* pro-cataal

Figure 2, which is the repetition of a part of figure 1, shows the mode of articulation of the double pieces *ee* and *oo* with the vertebral nuclei. The latter has two transverse processes, each of which arises from a fold of the tube, when the fluids that inflate it are removed. In the center there still exists a hole, the last vestige of the primitive form, that is to say, of the old tubular form of the vertebral nucleus.*

*Notochord. *Trans.*

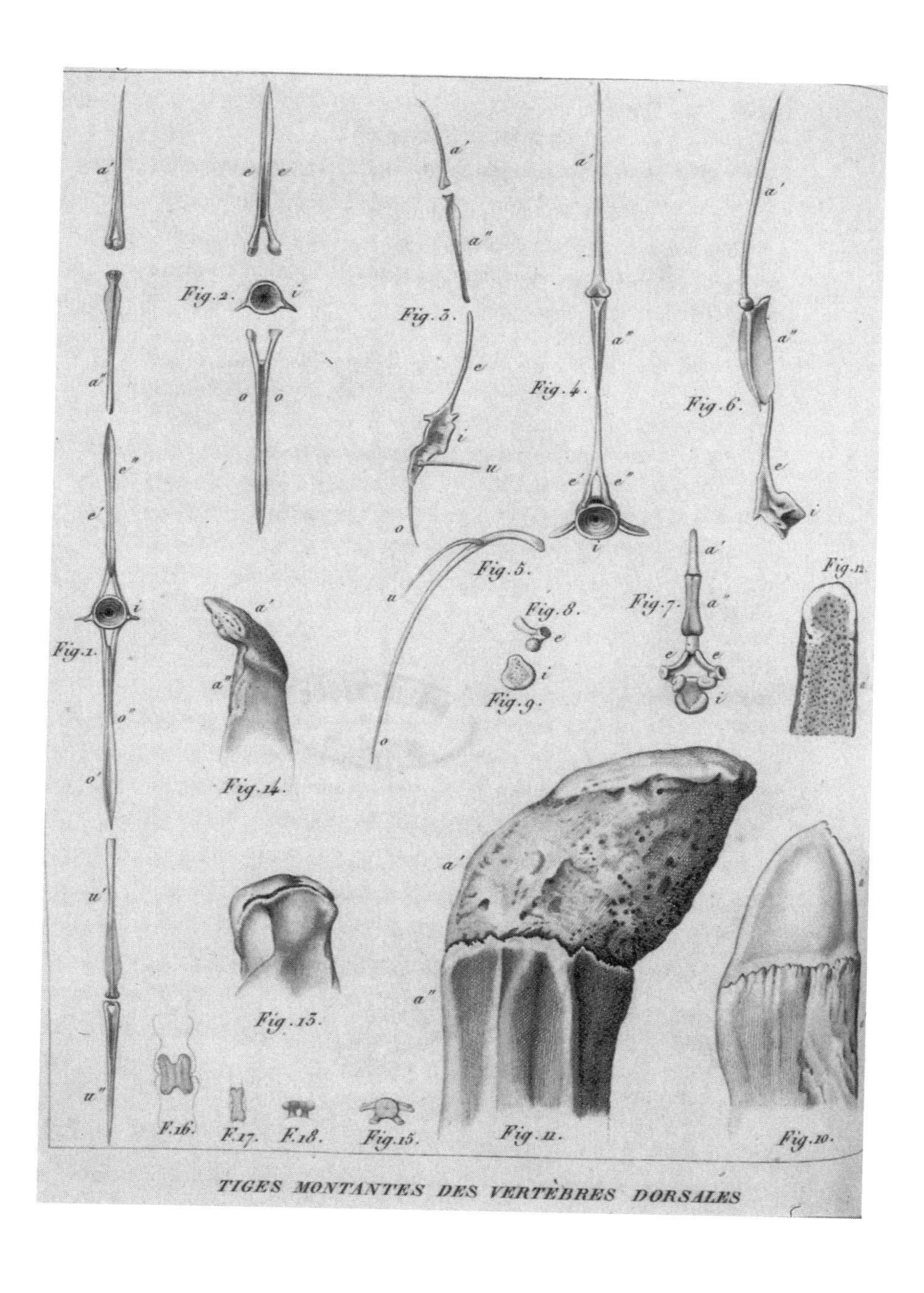

Plate V. The spinous process of the dorsal vertebra.

Figure 3 represents an abdominal vertebra of the same plaice, seen from the side. The vertebral nucleus has its transverse processes more elongated; at the bottom is the vertebral rib *o* or the bone called paraal: below is the sternal rib and the cataal, labeled *u*, a bone in a generally equivocal situation, always displaced by its articular membrane. The pieces *a′* pro-epial and *a″* in-epial do not differ in essential from what we have in figure 1.

Figures 4 and 5 repeat, with respect to the ordinary perch, *Perca fluviatilis*, the facts of the subject of figure 3; it is also an abdominal vertebra, but it is seen from the front. The intervertebral hole of the nucleus *i* is distinct. The sternal rib of the cataal *u*, rather than being carried wholly on the nucleus, is lower down, and is articulated with the paraal *o*, that is to say, with the vertebral rib.

Figure 6 shows the same pieces in profile, minus the vertebral and sternal ribs.

Figure 7 displays the same parts for a bovine fetus. The vertebral body *i* is entirely filled; this announces a development superior to that of the vertebrae of fishes that we have previously examined. The very solid perials *ee* were already, though feebly, attached to the nucleus. I have had these pieces drawn separately, figures 8 and 9. The in-epial *a″* would form an ossicle whose ossification was not terminated near the base. A cartilaginous part there bears witness to this state of imperfection. As to the part indicated by the letter *a′*, it would be entirely cartilaginous, and in that state it would be the matrix of another piece, the pro-epial.

This circumstance is established by the permanence of this fact in a calf that had been suckled for two months (see figure 10, letter *a″*), when, quite the opposite, the in-epial or the ossicle situated underneath finds itself, at that age, completed and already united with the other two dependencies of the vertebra.

The same demonstration is provided, a fortiori, by the facts of figure 11, where the pro-epial *a′* is seen in its state of complete ossification, but nevertheless in a degree to justify its being a separate bone, since at this moment the pro-epial is only articulated, but not yet fused with the in-epial.

Figure 12, representing a section of the ossicle *a′*, figure 11, furnishes another proof, which once more establishes that in fact the pro-epial, in the steer, is an isolated ossicle whose development is at first only delayed. The tissue of this piece is different from that of the other bones; it is denser in the interior; the compact part abounds, especially at the end of the bone, at the starting point of the ossification.

I wanted to find the corresponding facts in human anatomy, and I found, in the anatomical museum of the hospitals, an establishment created and directed by Dr. Serres, the two examples nos. 13 and 14; they represent the extremity of the spinous process (in man), to wit, in figure 13, of a dorsal vertebra, and, in figure 14, of a lumbar vertebra. This is without doubt only a pathological case, an extraordinary subdivision reverting to the conditions proper to animals.

We entertain a different idea of the normal state of these vertebrae. In fact, we have shown (p. 97), that in fishes, when the pieces of the lower girdle remain extraneous to one another at the partitioning of the vascular system, they are in an indecisive situation with respect to one another. In mammals, it is the same with regard to the pieces of the upper girdle in similar circumstances. The epials do not always occupy the same position at the top of the vertebra; they descend the length of the perials, one at the right and one at the left; they are called processes. I cannot provide a more explicit explanation in a memoir devoted to generalities, and I make this point in passing only to prevent its being thought that I infer nothing of consequence from the facts recorded in nos. 13 and 14.

The kangaroos, the bandicoots, and in general all the pouched animals, who help themselves with their tails to walk, to jump, and even to maintain themselves in an upright position, succeed in doing this by virtue of a relation of their postabdominal vertebrae to those of fishes. This relation consists in a more considerable development of those same vertebrae. Although in other mammals they are only rudimentary ossicles, gradually diminishing and finally disappearing, they are more heavily developed, and above all better provided with musculature. Then the bones that surround the vascular system appear to more advantage in the quadrupeds; then the same facts and the same relations as those to the rib cage reappear there in close to the volume. The paraals and the cataals exist below the vertebral bodies, but uniting and blending in one single piece, whose shape is determined by the objects interposed between its branches, that is to say, by the nutritive artery. This shape is usually that of a capital V, hence the name "V bone"* given to these pieces.

But in the pouched animals, in whom the tail is always in action, whether for locomotion, or even for the vertical position, these bones take on another shape; the acute angle of the V is replaced by a flat surface. These bones are arranged as a slab.

When further and further attenuated, they finish at zero toward the end of the tail, each branch of the V collapses, collides with itself; those bones first be-

*Or chevron bone. *Trans.*

come small nipples, then small molecular points. Anatomists have shown themselves unconcerned with them, admitting them as epiphyseal bones. However, these considerations make evident the following proposition. The smaller the four elements of the V bones become, the more those elements meet and penetrate one another, and the less their mechanical division can be visually witnessed. If, on the contrary, they grow larger, they are more than juxtaposed, and they appear distinctly.

M. Delalande, before his departure for the Cape of Good Hope, knowing that I was occupied with this research, gave all his attention to the V bones of cetaceans. The three skeletons of whales that we owe to his efforts lack only one of those ossicles, which are ordinarily neglected. Nine of the first, starting from the pelvis or the thirty-first vertebra, have the shape that has given rise to their original appellation. The pelvis itself is only an assemblage of these ossicles: there are two slender and elongated pieces (a paraal at the right and its congener at the left), from which hangs a large and strong plastron situated on the median line and arranged in a semicircle, that is to say, from which hang the two cataals welded together. The paraals of the next vertebra (the thirty-second starting from the skull, or the first of the coccygeal vertebrae) are also long rods free at each end: but these differ from the paraals of the pelvis, in that they are attached to the base of the cycleal, when the latter are attached to its sides. The nine V bones that follow are composed of a paraal on the right and the congener of that bone on the left, and further of a point of ossification at the place of their fusion, where the elements of the two cataals are blended. This circumstance is to be seen very distinctly in the V bones of a manatee skeleton in the anatomical collections of the Royal Garden: it is noticeable even in crocodiles. In the North Cape whale, reported by M. Delalande, the two paraals of the forty-second and forty-third vertebrae (the eleventh and twelfth of the coccygeal vertebrae) are isolated and parallel; the cataals that complete the belt extended around the artery remain in their primitive state of cartilaginous pieces. M. Delalande has also reported on the skeleton of a North Cape whale that had just been born. What I have just said about the last vertebrae of the adult whale are appropriate in all respects with regard to the first coccygeal vertebrae of the neonate: in the former it is a case of greater, and in the latter of less development. Thus it is established by these examples that the V bones are composed of four distinct pieces, most often fused together, but not always. I propose to call their union *furceal*, from *furca*.

In the kangaroos, the four units of the furceal (in the table here, V, otherwise almost everywhere) due to a superior development, although welded together, appear quite distinct: they are less so in the bandicoots, another genus of pouched animals. We have drawn four pieces of the tail of the latter; one taken among the small vertebrae, figure 15; one of the large bones, of those called a V, figure 10; another much narrower one, figure 17. These two ossicles are seen from below and longitudinally. This is a bone that is represented in number 18; but it shows its transversal and posterior face; its articulate branches are close to the point of touching and blending; one would say a second vertebra below the true one; but the hole in the center is nothing but the continuation of the tube traversed by the vessels.

Note: I do not include in the number of the nine units of the vertebra the bony plates interposed in young subjects between the bodies of the vertebrae. Those bones are to my eyes abortive vertebral bodies: there are other ossicles in the same situation, which tried to come into being, but have succeeded only imperfectly. Some time, in a special work, I shall deal with these bones, which owe their rudimentary condition to a tendency to abort.

PLATE VI. I beg my reader to pay attention to the conformation of the first figure. The use of the same letters makes evident the identity of the facts established in the present memoir, lowercase *i* of that first figure stands for the vertebral nucleus, but in the insects that nucleus never closes and remains as a tube, while in fishes it is all that remains of the vestiges of that primitive organization. The letters *a′*, *a″*, *e′*, *e″* express the fact that the bones they designate are upper pieces, as the letters *o′*, *o″*, *u′*, *u″* indicate the lower ones. These last are bounded by cutaneous appendages.

We shall not return to figure 2; we have given a sufficient explanation in our text.

Figure 3 represents one of the rings of the thorax; one can see there that the cycleal no longer forms a closed ring of all the parts. It takes up to seven pieces to form these appendages. Entomologists have reduced them to a fourfold division—the haunch, the thigh, the leg, and the tarsus—; but then two pieces are counted as forming one of these subdivisions.

PLATE VII. In this plate, we shall refer for the moment only to figure 2; it is a longitudinal section of the lobster. We have turned the animal inside out and have placed its abdomen above and its back below. In this situation the organic systems are arranged as in mammals: *a* is the spinal marrow, *b* the dorsal muscles, *c* the intestinal canal, *d* the heart, *e* the aorta, *f* the carotid arteries, *g* the pulmonary arteries, and *h* the intestinal muscles.

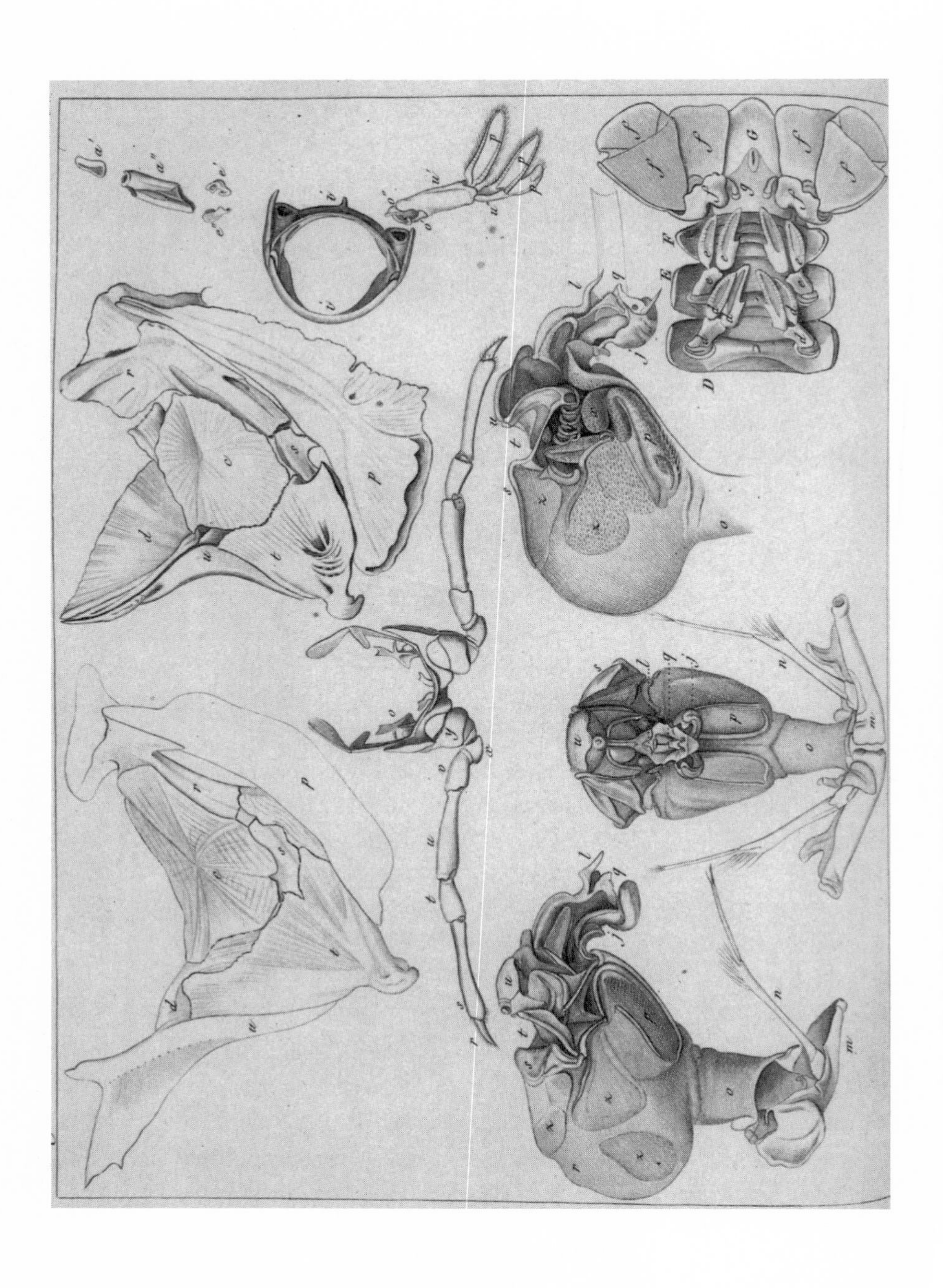

Plate VI. The anatomy of the lobster. 1.

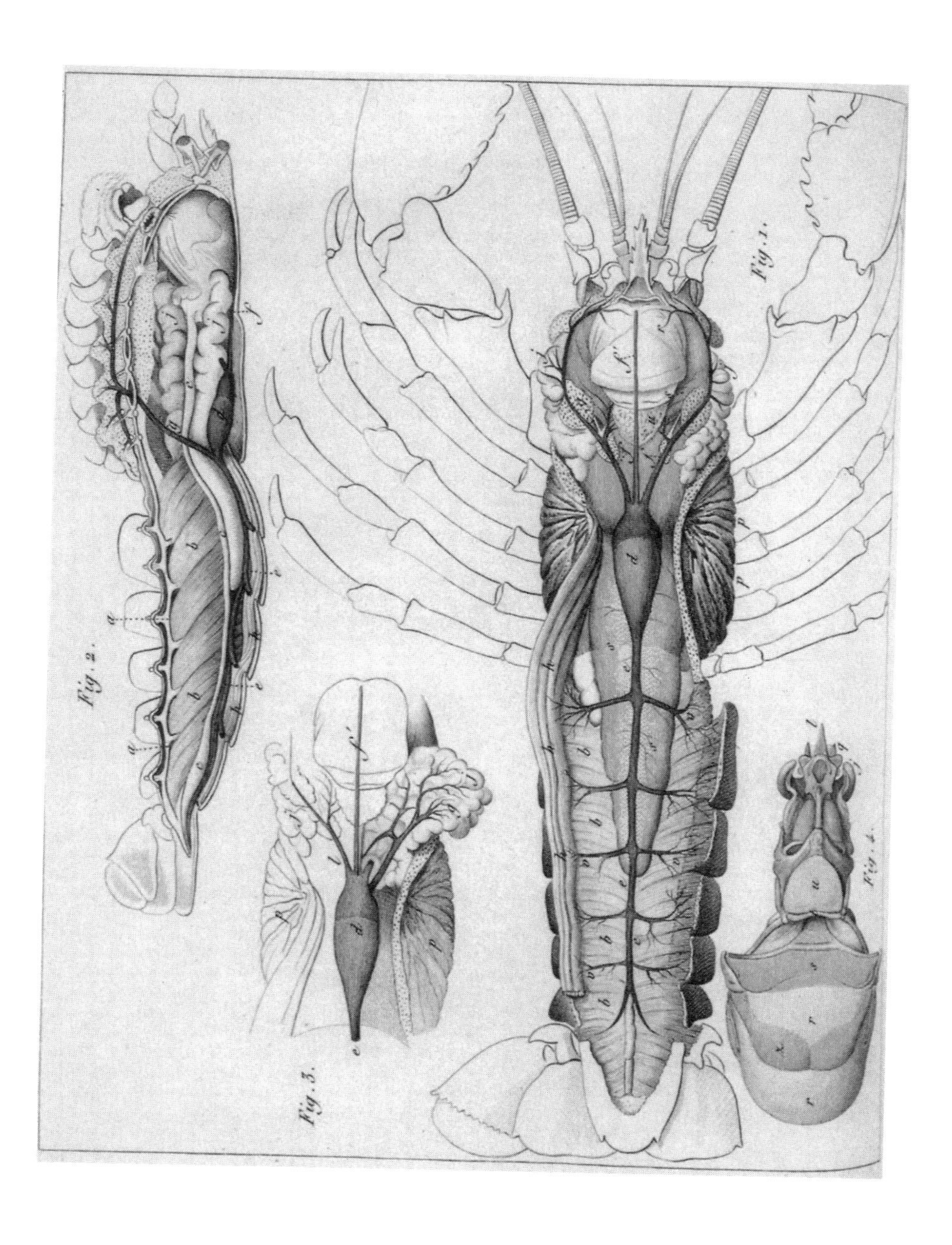

Plate VII. The anatomy of the lobster. 2.

4

THE AFFAIR OF THE CROCODILES OF NORMANDY

The affair of the fossil crocodiles of Caen and Honfleur is seldom presented with all the attention it deserves. Nevertheless, it is an essential part of our story, since it begins in 1825 and is taken up again in the autumn of 1830, after the controversy at the Academy of Sciences. Further, it is not without interest: this time it is really a question of evolution.

The second edition of Cuvier's Investigations of Fossil Bones appeared in 1825.[1] It is considerably larger than the edition of 1812. In particular, there is an important chapter in volume 5: "On the Fossil Bones of Crocodiles." Cuvier's approach to the matter is surprising, for in the reproach that he addresses to his colleague Faujas, we are shown all the difficulty of the determination of fossil fragments:

> The fossil crocodiles appear not to be very rare in the old secondary strata, and what is remarkable is that although they belong to species different enough from one another, they almost all belong to the subgenus with an elongated snout that we have named the subgenus of the gavials. A superficial examination had even led the late M. Faujas, on the assurance of Merck, that they belonged to the true gavials of the Ganges.

From this preamble we see that Cuvier wanted to give his reader an a priori idea of these crocodiles, by announcing at the start that the fossils he is about to describe are all gavials, that is to say, animals that exist at present only in other climatic conditions. The only problem for Cuvier is no longer to examine and to classify the fossil animals of Normandy, but rather to explain a change of climate. He does not yet know that in doing this he will encounter grave problems coming from his colleague Geoffroy Saint-Hilaire.

Let us go back a few years. In 1817, a former student of Lamarck's described in the Annales générales des Sciences Physiques a piece found by Jacques Armand Eudes Deslongchamps (1794–1867) in the quarry of a village called Allemagne, near Caen.[2] In 1822 he found another fossil coming from a nearby quarry, the quarry of Quilly. After long palavers, Cuvier succeeded in having the rocks sent to the museum. He described them minutely, bone by bone, particularly the skulls. Cuvier's formulation is curious and lacks the rigor that is usually evident in his work. Read what he says:

> The first glance given to this piece thus freed from its gangue announces that it had belonged to a gavial, differing as much from those now living and from the fossil forms discovered so far.

Then later:

> We see according to the imprint, figure 13, that the snout of this crocodile was still longer in proportion than that of the gavial.

Finally:

> It [the head] grows more slender in front, like the impression it has left on the rock. Its depression is a bit stronger than for the gavial; the bones of the neck descend further down and form a more acute angle.

Despite his hesitations, Cuvier maintains his first diagnosis: it is a gavial.

Cuvier's embarrassment is even more marked in the description of other animals probably coming from secondary—liassic—Norman terrains at Honfleur and Le Havre. Thus, while article 4 of his chapter has as its title: "On the Bones of Two Unknown Species of Gavials, Found Pell-mell near Honfleur and Le Havre," we find, several lines later:

> It offers . . . more characters than are needed to determine a species and to differentiate it from the gavial, and it does not resemble it sufficiently, so that I can base on that snout a sufficiently probable conjecture concerning the animal from which it comes.

Further along, a propos of another species with a long snout, we read:

> It results from this that the skull of the fossil has an oblong form quite different from that of the gavial; it is joined to the snout by an imperceptible narrowing, and not by an abrupt contraction.

Despite this shilly-shallying, Cuvier nevertheless concludes that these are two different "species" but belong to the genus gavial.

How does it happen that an anatomist as established and as conscientious as Cuvier can suddenly be so imprecise, so little assured in his descriptions? Here he presents an awkward reflection, hesitating on the characters to be considered, relying on a vague idea of resemblance. We must wait for the conclusion of this article 4 to have the solution:

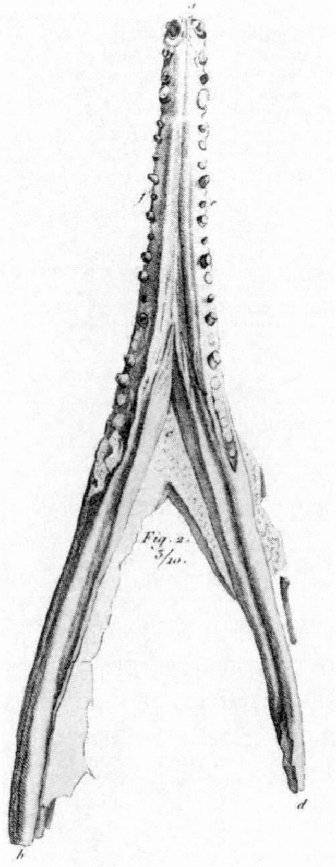

> Moreover, I believe I need no longer insist on the distinction to be established between these crocodiles and those that we know as living forms, or be obliged to reply to those who still those that the extraordinary difference that distinguish these two kinds of bones from those of the gavial by the influence of age, by nourishment, by climate, or by passage to the state of petrification, as the late M. Faujas wished to do in the passage cited at the beginning of this chapter. Would all the other causes together have been able to place at the front the convexity that the other crocodiles have behind their vertebrae? Could they have changed the origin of the transverse processes, flattened the edges of the orbits, decreased or

The lower jaw of the crocodile of Honfleur, as copied from a plate in Cuvier's book, resembles that of an extant gavial.

increased the number of teeth, and so on? That is as much as to say that all our extant species come from one another.

The last phrase of this passage is crucial; it shows us that Cuvier was afraid of bringing arguments to the enthusiasts for evolution, represented in fact at that time, as he saw it, chiefly by Lamarck, whom he had been combating for twenty years.[3] This example illustrates well to what point the determination of a fossil—which may seem banal to a nonbiologist—can have conceptual repercussions of capital importance. And Cuvier knew his subject too well not to have immediately understood this. So he refused comprehensively in order not to give an entrance to the idea of transformism. Would it be imprudent to postulate that it was reading that conclusion that motivated Geoffroy Saint-Hilaire to take up again the study of these alleged gavials? Let us not forget that, ever since his voyage to Egypt, he had become a veritable specialist on crocodiles. "I had an exquisite feeling for the species of crocodiles," he said.

Some time earlier there had appeared—at last!—the end of the proceedings of the Egyptian voyage of the team of scholars who had accompanied Bonaparte. To be sure, various reports and illustrations had been assembled with much effort, each person wishing to publish his results separately. But most of the work had been accomplished after the fall of the Empire. For obvious technical and political reasons, the process of publication had been protracted until 1825. Most of the description of reptiles was edited, under his

Ever since the Egyptian expedition, Geoffroy Saint-Hilaire had been a great expert on crocodiles. Here we present the corresponding plate from the Description d'Egypte. *Above: the common crocodile and its young. Below, left: the spiniped stellion; right: the stellion of the ancients.*

father's supervision, by Isidore, Etienne's son, who was then an assistant at the Museum. The description of the crocodiles occupies a large part of the volume. In particular, Geoffroy Saint-Hilaire here studies in great detail the nasal canal, which seems to him worthy of interest, as much for its physiological role—it serves to store air when the animal dives—as for its special anatomical characteristics, which permit crocodiles to be distinguished from mammals. In particular, Geoffroy Saint-Hilaire lingers over the curious position of a bone that he calls "herisseal"—technically the internal pterigoid process of the sphenoid—, of which he gives the embryological origin. In several pages he explains its importance, chiefly in regard to the soft palate. Thus it is not lightly that Geoffroy Saint-Hilaire claims to have found, using the principle of connections, an anatomical character in which "resides the cranial essence of the crocodile." It is easy to see why he hurried to examine more closely the fossils that had just arrived at the Museum.

The resulting memoir, entitled "Investigations into the organization of the gavials," published in 1825, is explosive, as we can see from the flowing subtitle: "On the natural affinities from which results the necessity of a new generic distribution, Gavialis, Teleosaurus et Steneosaurus; and on the ques-

The Nile lungfish, or bichir, is one of the new fishes discovered by Geoffroy Saint-Hilaire. This fish is so unusual that Cuvier would say its discovery alone justified the expedition to Egypt.

tion whether the Gavials (Gavialis), widespread today in the oriental pther from the fossil Gavials called "crocodiles of Caen" (Teleosaurus) or from the fossil Gavials of Le Havre and Honfleur (Steneosaurus).[4]

Explosive, since this was really a double slap at Cuvier. To begin with, Geoffroy Saint-Hilaire was flaunting an error in classification—and hence in comparative anatomy—on the part of his colleague. But above all this was to maintain the thesis of evolution, and to claim that there was a living continuity—and not by successive creations, as Cuvier lets it be understood—, but also that in the course of that descent species could be transformed. In this manner, by this groundbreaking memoir, Geoffroy Saint-Hilaire was establishing the foundations of evolutionary paleontology.

In fact, Geoffroy Saint-Hilaire had quite simply the stupefaction of seeing, in the skull of the crocodile of Caen, a form of herisseal that, for him, was characteristic of mammals. Thus he baptized the animal Teleosaurus and considered it "as if it were a mixed product of mammal and crocodile." For him, following his style of thought, the Teleosaurus was "in advance" of extant crocodiles. As for the animals from Le Havre and Honfleur, like those of Quilly, following the same schema they would be "behind" the extant forms. That is why Geoffroy Saint-Hilaire called them Steneosaurus. Thus he built up intermediaries: "To show all these animals in natural order, we shall say of the reptiles of Honfleur, that they must follow the genus Crocodilia at some distance, and of Teleosaurus that it must immediately precede this genus."

However, Geoffroy Saint-Hilaire had not previously believed in evolution, in the transformism proposed by his colleague Lamarck, although he loved and respected him. Unlike Cuvier, he had never attacked him; but he had never supported him, which indicates that on the whole he was critical of him. This study seemed to him like a revelation. In fact he was now studying "the degree of probability that the Teleosauruses and the Steneosauruses, animals of antediluvian times, are the source of the crocodiles prevalent today in the hot climates of two continents." He proposed an evolution governed by changes in physical conditions on the globe, and in this connection he cited the laws of Lamarck; in a moving note he recommended reading them, as if he wished to be forgiven for having been blind too long.[5]

But Geoffroy Saint-Hilaire went quickly to work, and outstripping Lamarck, he at once incorporated the whole of his knowledge, in particular in embryology. He then suggested that the mechanism of the formation of monsters, which he then knew very well, might be an analogue of an evolutionary mechanism:

> What demands considerable time in the great operations of nature is nevertheless accessible to our senses and occurs in miniature and under our eyes in the spectacle of monstrosities, whether accidentally or voluntarily produced. . . . Thus beings that would be different from their ancestors could, at the instance of monstrosities in their maternal stock, arise from those ancient sources: from which, urging the consequences of these truly incontestable facts, I can propose that it is by no means repugnant to reason, that is, to physiological principles, that the crocodiles of the present epoch might descend by a succession without interruption from antediluvian species, recovered today in the fossil state in our territory.

Cuvier did not hesitate to counterattack in the article "Nature" in the Dictionnaire des sciences naturelles (Dictionary of the Natural Sciences), volume thirty-four of which was published in the same year, 1825. He did not defend his crocodiles, whom he perhaps thought indefensible, but placed himself on a philosophical and religious plane, repeating his criticism of the concept of the unity of organic composition:

> We conceive of Nature simply as a production of the all-powerful governed by a wisdom whose laws we discover only through observation; but we think that these laws relate only to the conservation and harmony of the whole; that, consequently, everything must be well constituted in such a manner as to concur in that conservation and that harmony, but we do not perceive any necessity for a Scale of Beings, or for a unity of composition.[6]

Geoffroy Saint-Hilaire was deeply affected by that article.

In fact, Geoffroy Saint-Hilaire's thought is not as close as is believed to current transformism. The idea of an intermediary came to him starting with the moment at which he could make use of the principle of the balancement of organs, as in embryology and comparative anatomy. In contrast to what Darwin would later propose, it is not so much the succession of generations in interaction with the environment that Geoffroy Saint-Hilaire envisaged, but the evolution—in the original sense of that term—, the unfolding of a predetermined plan of organization that becomes step by step more complex in the course of time. Moreover, it is this idea that Etienne Serres, a convinced antievolutionist, was to develop at the funeral of his friend and master.

This also explains why, in 1826, Geoffroy Saint-Hilaire would return to experimental embryology. He was seeking "to draw organization into unaccustomed paths," and, among the factors capable of influencing embryological development, he placed at the head the quality of the environment, chiefly that of the respirable air. He was trying to obtain a teratogenesis through modification of the external environment of the egg, thus attempting to justify experimentally his explicative hypothesis of animal evolution: To repeat: "What demands considerable time in the great operations of nature is nevertheless accessible to our senses and occurs in miniature and under our eyes in the spectacle of monstrosities, whether accidentally or voluntarily produced." Thus he wanted to imitate artificially the natural process of the genesis of monstrosities, in order to understand them better. By the same token, he wanted to show that environmental perturbations induced variations in development, thus bringing into existence what Jean Rostand would call "experimental transformism."[7]

It is clear that through modifications of the environment Geoffroy believed he could keep certain hens' eggs at the reptile stage. We can understand why Cuvier was uncomfortable, he who proclaimed that if even a single species would be successfully modified, geology itself could be touched in one of its fundamental principles.[8] He even tried several times to have Geoffroy Saint-Hilaire's experiments forbidden by the police, because contrary to religion, he was using for "scientific" purposes his position at the Ministry of the Interior. This action is also of the highest interest, since it suggests that Cuvier was far from being sure that such experiments would necessarily fail.

The decade was marked by various skirmishes. Cuvier, in the first volume of his Natural History of Fishes, declared that the doctrine of the unity of composition "has reality only in the imagination of some naturalists, more poets than observers." In 1829, Geoffroy Saint-Hilaire once more evoked, in the light of embryology, the relations between extant and fossil species.[9] Thus the affair of the crocodiles of Normandy led to the crystallization of a violent antagonism which had had its origin in the Memoirs on Insects.[10] Everything was in place for the great debate.

5

THE PUBLIC CONFRONTATION

The controversy of 1830 was to break out when, starting from the conclusions of a memoir submitted to the Academy of Sciences by Laurencet and Meyranx, Geoffroy Saint-Hilaire, in a grand unifying dream, tried to attach to the couple vertebrate/insect—or rather, by then, higher vertebrates and dermal vertebrates—the embranchement of mollusks, an embranchement that had been sacrosanct for Cuvier ever since his early years in Normandy. Clearly it is the unity of organic composition that is again in question. But let us take the facts in their chronological order.

In 1823, Laurencet and Pierre Stanislas Meyranx[1] had already presented a first—well-received—work on the cephalopods, the class of mollusks that includes nautiluses, cuttlefish, squid, and octopuses. In October 1829, they submitted for examination at the Academy of Sciences a new memoir entitled *Quelques considérations sur l'organisation des Mollusques (Some Considerations on the Organization of Mollusks)*.[2] Four months later, the two authors had still had no reply. It seems probable that Cuvier, the secretary of the Academy, might have been the cause of such a delay. They wrote to the Academy, and this time Geoffroy Saint-Hilaire and Latreille were named as referees. On February 9 they went to work, made dissections along with the two authors, and composed a report that would be presented in the course of the session of the Academy on February 15, 1830.

Laurencet and Meyranx wanted to show that, with a torsion—a plicature, in their terminology[3]—the organs of a cuttlefish present a relative arrangement identical to that described in vertebrates. For their demonstration they used the principle of connections. The comparison was pushed very far, indeed, too far. But Geoffroy Saint-Hilaire was clearly enthusiastic. He was offered what he had not dared to dream of: the possibility of conceiving a unity of plan between vertebrates and mollusks. Thus three of Cuvier's four embranchements would find themselves unified.

In all probability Geoffroy Saint-Hilaire was the only author of the report, which was clearly very flattering. No one would be surprised to find there an appeal to "the universal law of nature, the unity of organic composition." In fact, the conclusion of the report presented to the Academy on February 15, 1830, exceeded—by far—that of the authors: "MM. Laurencet and Meyranx have understood how to appreciate the needs of science, since they have tried to lessen the hiatus observed between the cephalopods and higher animals." Latreille, though a cosigner, disengaged himself at once, and wrote to Cuvier: "Be persuaded, my dear colleague, that I had no part in this publicity." Meyranx, much embarrassed, tried to remain in Cuvier's good graces: "We were far from believing that, on the occasion of a single simple consideration of the organization of mollusks, it would be possible to draw such exaggerated conclusions."

This time the disagreement became public, and Cuvier, directly implicated in the report, was in a situation such that he felt himself obliged to respond. For the publication of the proceedings, he obtained the suppression of the pamphlet in which he had been implicated (nevertheless, this pamphlet would be published in the work of Geoffroy Saint-Hilaire that will be found below). Then he decided to argue publicly, and the following Monday, at the session of February 22, he presented a note: *Considération sur les Mollusques, et en particulier sur les Céphalopodes* (a summary of it will be found in the work of Geoffroy Saint-Hilaire, following an article by a student of Cuvier's published in the *Journal des débats*). The celebrated anatomist's argumentation is interesting. First, he insists on the difficulties of vocabulary, on the fuzziness cloaking the words *composition* and *plan;*[4] then he tries to reduce the scope of his colleague's ideas, which could serve only to point out analogies. Looked at in this way, Geoffroy Saint-Hilaire would be no more than a simple continuer of Aristotle. Finally, citing some successes of his colleague that he had formerly applauded, Cuvier concluded that "he has done nothing to add to the ancient and familiar foundations of zoology."

Geoffroy Saint-Hilaire immediately tried to improvise a reply, which basically added nothing. He was visibly disturbed by Cuvier's attack, although, paradoxically, it did not criticize Laurencet and Meyranx's memoir in detail, nor destroy entirely the principle of a single plan, but tried to subordinate structure to function, privileging Cuvier's own principle of correlations. At the end of this stormy session, various zoologists offered their services to act as intermediaries between the two old friends. But Geoffroy Saint-Hilaire could not endure Cuvier's reducing his original idea to simple corollaries of a functional principle.

The following Monday, at the session of March 1, instead of replying, Geoffroy Saint-Hilaire read a memoir—"De la théorie des analogues, pour établir sa nouveauté comme doctrine, et son utilité pratique comme instrument" ("On the theory of analogues, to establish its novelty as doctrine and its practical utility as an instrument of research"), in which he presented his ideas in synthetic fashion, retracing their genesis. In the main text, he found himself forced to argue in a precise manner for the interest of his theory. Thus we find here the basic elements of his theory, matured and expressed in a new way:

> I received an impression which, returning identically time and again, led me to entertain this view, namely, that so many animals that I considered different and which I treated as distinct by giving them a specific name, nevertheless did not differ except for some trivial attributes that modified more or less a structure that was generally and evidently the same. . . . Faced with cabinets of ornithology, I noticed on the shelves only the repetition, multiplied a great number of times, of the type *bird*.

We find here the notion of *type* that will later be taken up by Owen, and which corresponds to that search for the invariant dear to Geoffroy Saint-Hilaire:

> I distinguished only general traits, to wit, the head, the neck, the trunk, the tail, the wings, the feet; in all the individuals, there were

---

To refute the argument of Laurencet and Meyranx, Cuvier, in his article "Considérations sur les mollusques et en particulier les Céphalopodes," represents a quadruped curved like a U, plate A, and an octopus, plate B.

The figure A represents the section of a quadruped, folded over on itself, so that the pelvis meets the nape of the neck.

The figure B is the section of an octopus in its natural state.

In the two figures, the same letters indicate the same organs: a, a, *the brain;* b, b, *the ear;* c, c, *the upper jaw of mandible;* d, d, *the lower;* e, e′, e″, *the esophagus, the stomach, and the intestinal canal;* f, f, *the liver;* g, g, *the respiratory organ (lung in the mammal, gill in the cephalopod);* h, h, *the aortic heart;* k, k, *the principal vein;* l, l, *the principal artery;* m, m, *the organ of generation;* n, *the kidney, peculiar to the mammal;* o, *the bladder, again peculiar to the mammal;* p, *the funnel, peculiar to the cephalopod;* r,r, *the anus;* s, s, *the orifice of the genital organs;* t, t, *the spinal cord, peculiar to the mammal.*

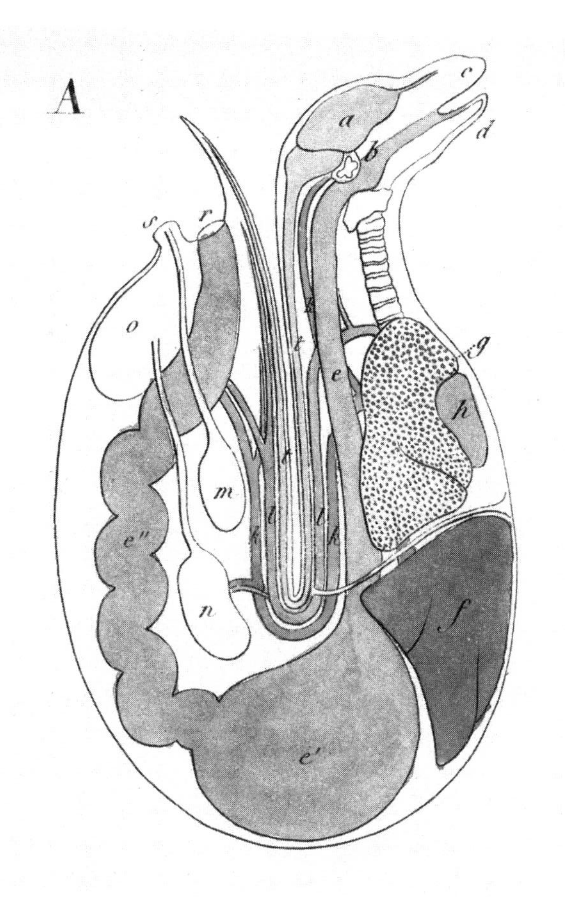
A

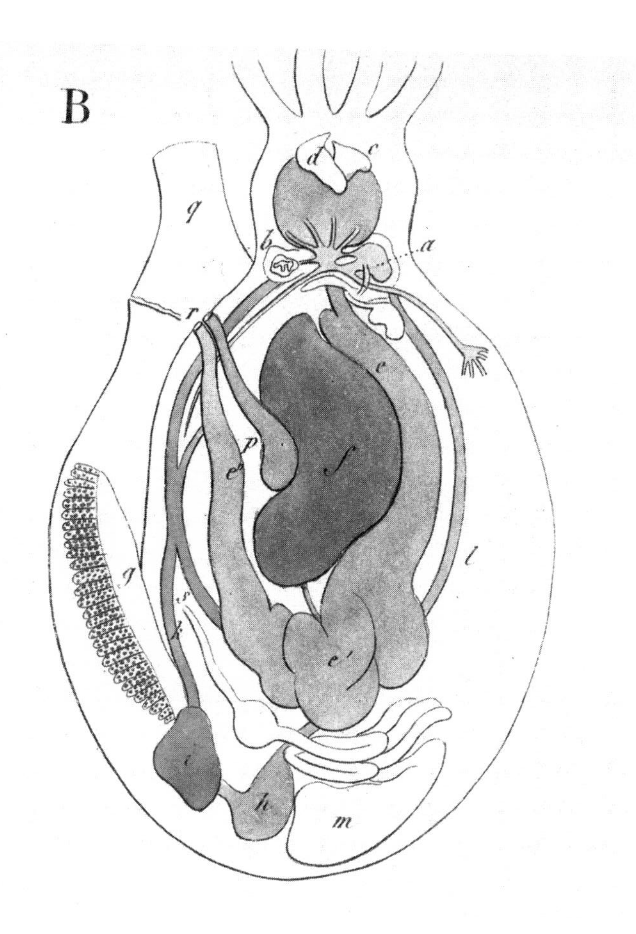
B

feathers as teguments; in all, a horny beak surrounding the jaws: all these things repeated exactly, and, further, existing in places respectively the same.

Naturally, there follows the principle of connections, which "served me as a compass and kept me from error in the *investigation* of identical materials."

To illustrate his remarks, Geoffroy Saint-Hilaire took as an example the hyoid of vertebrates, which he had studied in the first volume of his *Anatomical Philosophy*. Technically, he showed that his principle led to the correct interpretation of the styloid process. In conclusion, he recalled the application of his theory to the study of monstrosities, on this subject taking up again the theme of the second volume of the *Anatomical Philosophy*.

Geoffroy Saint-Hilaire then fell ill. The discussion was resumed only at the session of March 22, in the course of which Cuvier read, in reply, a memoir entitled "Considerations sur le os hyoïde" ("Considerations on the Hyoid Bone"). In this text, a veritable indictment, he followed step by step the study made by Geoffroy Saint-Hilaire, showing that each bone changed connections as the species changed, and that therefore the principle of the unity of plan could not be applied to it. He then repeated the analysis in such a way as to apply his own ideas to the classification. The approach taken here by Cuvier is quite strange. He attacks while confining himself to the vertebrates, an embranchement for which there seemed to be a consensus of opinion. This strategy might seem ingenious since, if Geoffroy Saint-Hilaire's concepts could not be applied here, all generalization would be a useless effort. However, in proceeding in this way, he was doing himself a disservice, given that he was undermining the idea of the existence of a plan for each embranchement. After this violent diatribe, Geoffroy Saint-Hilaire reserved his reply on the hyoid for the session of March 29 and proposed the reading of a memoir he had prepared: "De la Théorie des analogues, appliquée à la connaissance de l'organisation des poissons" ("On the Theory of Analogies, Applied to the Knowledge of the Organization of Fishes"). This text is extremely valuable for the understanding of Geoffroy Saint-Hilaire's transformism. In particular, we find here what he means by "unity of system in the composition and arrangement of organic parts."

He begins by explaining why animals very far apart in the scale of complexity cannot with impunity be compared with one another. His reasoning cannot be understood unless we accept the idea of the "career" of the embryo, which we discussed earlier. In view of the different degrees of organization, the comparison mollusk/vertebrate can be seen as a comparison of the adult with the embryonic state:

> This is equivalent to considering the beings that belong to one of the ages of the variable developments of organization. And in fact it is correct to consider the mollusks as realizing forever one of the inferior degrees of the progressive order of organic developments, to see them as having stopped at this point, and because of this fact, as not having yet furnished a certain kind of organ, or if any such has begun to appear, as not yet having produced it in full force.

This last phrase is extremely important; it proves that Geoffroy Saint-Hilaire does not conceive of a "plan" only with the help of organs that are observed in dissections, but also by taking account of all its potentialities, even if they have not yet been expressed. Thus Geoffroy Saint-Hilaire looks at things in two ways: that of the eyes of the body, which observe during a dissection, and that of the eyes of the mind, which interpret organization in the field of transcendental anatomy. Thus a "system" is the organic totality formed by all the actual organs and all their *potentialities.*

In the last analysis, the deepest element in the thought of Geoffroy-Saint-Hilaire consists in his manner of thinking of temporality starting from the consideration of organic series:

> Thus there would be, there is truly but one system of organic composition, only one primitive design to govern the arrangement of its parts, only one plan, finally, unvarying with respect to what forms the essence and the intertwining of the elements comprised in all organic formation.

This vision is in fact close to that of Kielmeyer or of Lorenz Oken (1789–1851), and of all the German zoology inspired by the Philosophy of Nature.[5]

On March 29, Geoffroy Saint-Hilaire replied to Cuvier's argument with his memoir *Sur les os hyoïdes* (*On the Hyoid Bones*). His defense has little structure, and gives the impression that he did not know how to organize it. He begins by explaining that if the different hyoids have the same name, that may be because they are of the same nature. . . . Then, to answer one of Cuvier's criticisms, he returns to his work of 1818, in which he had shown that the number of bony pieces of an organ may vary. Then we arrive at the most interesting point of his argument—and perhaps the most enigmatic—, when he applies his highly personal notion of system to animals lacking a hyoid:

> The moment, the suitable age, is needed in any embryo, of man, of mammal, of bird, etc., for the hyoid to appear: earlier it is not compatible with the degree of organization of that period. In the same way as in animals that belong to that same degree of organic development, there is not, there cannot be, a hyoid; what is surprising about that?

We see here that Geoffroy Saint-Hilaire is assimilating completely embryology and evolution. More precisely, for him, a present-day animal that does not present a hyoid on dissection possesses it *virtually*.[6] In fact, if the embryo had not been stopped in its "career," it would have developed a hyoid at a "superior" stage of development. This reveals the chasm that separates the ideas of Geoffroy Saint-Hilaire from those of Darwin. The latter wanted to show that there is no preestablished path and that the choice between possible futures is realized in the dynamics of each instant.

There were considerable public repercussions of the controversy. The academicians, who did not like the backwash, tried to stop the quarrel. Further, Anaïs, one of the daughters of Geoffroy Saint-Hilaire, died at this juncture. Cuvier, who had lost his daughter Clémentine two years earlier, went to grieve with him. Thus the personal bonds held despite the scientific dispute. Overtaken by these events, the two protagonists agreed to stop their public controversy, though they were to resume the consideration of their difference later when things had calmed down.

Four months later came the Three Glorious Days, the revolution of 1830. Cuvier went into exile for some months in England. On his return he found a radically altered political and intellectual environment.

And evolution?

As we have seen, this controversy can be considered the culmination of the succession of disagreements that were approaching a crescendo starting in 1820. At first the various skirmishes seem to have had different major themes, like classification, embryology, paleontology, and evolution. That is why studies lacking inclusiveness do not offer a consensus on the immediate cause of the controversy of 1830. One author sees there the struggle of evolutionism against fixism, another reads in it the opposition between catastrophism and gradualism, a third puts first the influence of the revolutionary ambience of the period. . . . However, from 1820 on, all the debates present as their common denominator the principle of the unity of organic composition. In fact the key question is always the same: can the resemblance of plan that the scientific community of the time agreed to recognize *within* each embranchement be extended *between* embranchements? This truly novel vision of the animal world entails a change of perspective for many other biological themes. At the same time, such a challenge bears on certain domains of philosophy, as well as the social sciences.

Nevertheless, it is evident that the question of evolution underlies the debate. That is why, for the sake of clarity, it is appropriate to reallocate the position adopted by Geoffroy Saint-Hilaire with respect to the basic question, doubtless related to the debate. Let us give a rapid accounting of the situation before February 1830. Geoffroy Saint-Hilaire demonstrates the unity of plan among the vertebrates, using the stages of embryonic development. A pause; then he unites vertebrates and insects, with criticisms from Cuvier. Next, in order to confirm definitively the pertinence of his method, he turns to the interpretation of monstrosities, then to that of fossils—a domain that had been reserved for Cuvier.

The principle of connections, a conspicuously geometrical principle, is the compass that allows him to take his bearings in comparative anatomy as well as in embryology. He uses it to classify monsters, going so far as to compare the monstrous state of one species with the normal state of another species close to it. Finally, Geoffroy Saint-Hilaire and Serres attach to the totality of living things the succession of stages that they reveal in beginning with embryonic development, but always in a static manner. Thus they establish a strong relation between the parallel description of organisms in zoology and a sequential description of the different stages of embryogenesis, playing a game of musical chairs between the synchronic and the diachronic, the permanent and the transitory:[7]

> First we see the transitory form of higher embryos assume fugitively and in passing the organic and permanent attributes of lower animals; further, the permanent organization of the latter sketches in its successive degrees of perfection all the embryonic phases of the one among them that most closely approaches the last of the vertebrates in such a way that, for them as well, the various features of their zoology are only in some way the graduated scale of their organogenesis.[8]

But let us not deceive ourselves: the time that appears necessary to pass from one zoological form to another is not, at that period, directly that of paleontology, a rhythm produced by the succession of generations. We must await Darwin to have that vision of things.

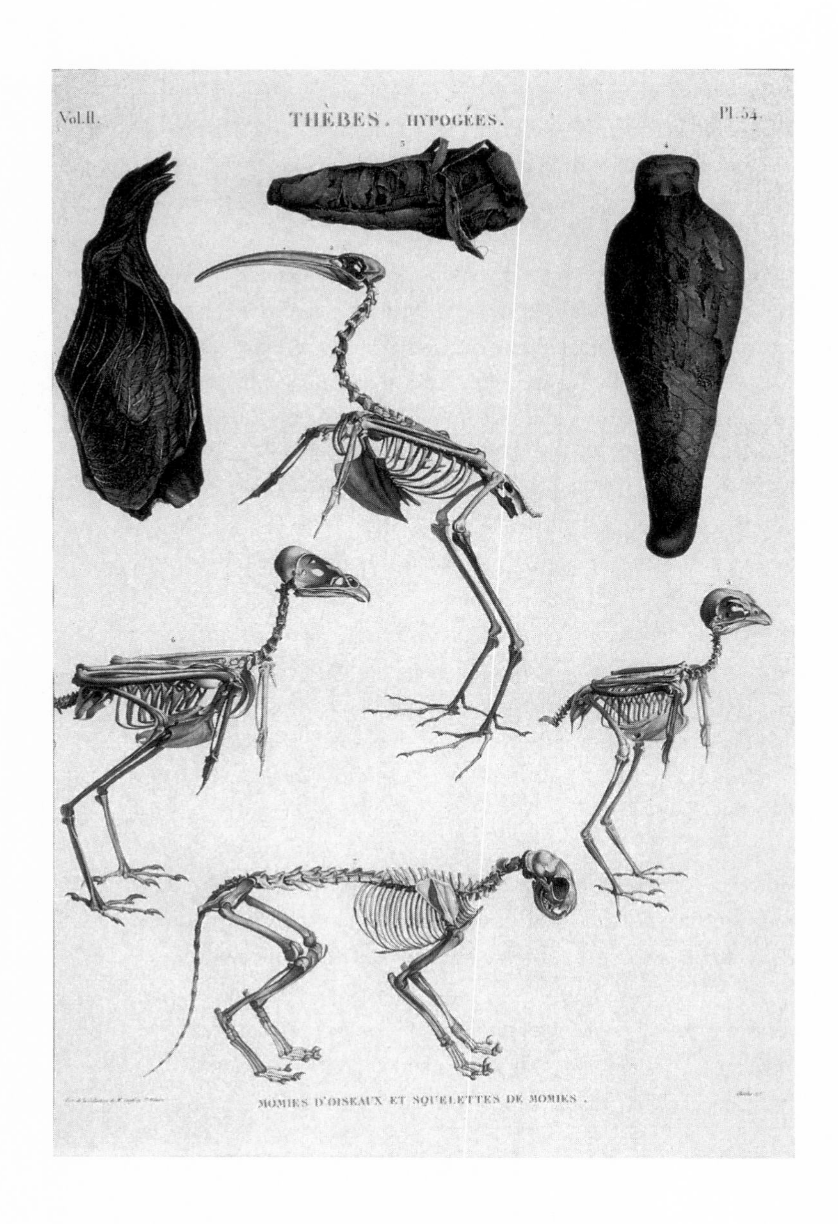

Geoffroy Saint-Hilaire brought back animal mummies from Egypt, in particular those of the ibis (at the center of the plate). Cuvier dissected them, showed that their skeletons were identical to those of extant forms, and considered this fact to falsify transformism. The plate is taken from Description de l'Egypte.

It was this idea that Serres chose to present in the course of the eulogy that he pronounced at the funeral of Geoffroy Saint-Hilaire (see the complete text in the collection of funeral orations that follow chapter 6):

> In short, science reveals that progressive and continuous march of life marked at long intervals by halts that seem to be for nature times of rest. Thus does the whole animal kingdom appear as but a single being that, during its formation, stops in its development, here sooner, there later, and thus determines, at each interruption, the distinctive characters of classes, families, genera, and species.

This law of *recapitulation*—which the history of science will retain as the law of Geoffroy Saint-Hilaire-Serres-Meckel—presents a finalistic conception of the succession of living forms on earth.[9] The order is preestablished; the potentialities are given, as if it were a case of embryogenesis. For Serres and Geoffroy Saint-Hilaire, certain possibilities present themselves at every instant to an embryo in process of formation, in the form of pathways, of obligatory and preexisting turning points that generate a multiplicity of paths, each of which leads to an animal. To their minds, these stages—these choices—do not necessarily reflect moments of the past of the species. Remember that Geoffroy Saint-Hilaire was a mineralogist by training, the favorite student of Haüy. Thus he would look for invariants in the composition of organisms, as Haüy looked for invariants in the structure of crystals. The notion of unity was his idée fixe, but the ultimate foundations of that unity did not enter into the field of his inquiry.[10]

Thus, following Serres and Geoffroy Saint-Hilaire, one and the same "career" including the whole hierarchy of possible forms lies open for the genesis of animals.[11] They stop earlier or less early, following a scheme that suggests a hierarchy of perfection, a new Scale of Beings which, as distinct from the older one, is no longer linear, but bushy. With this logic, there are again "higher" and "lower" animals. But it is not the past of the species, transmitted in hereditary fashion, that dictates to the embryo the paths of its differentiation. On the contrary, it is the immanent series of stages of progression, actualized by the order of embryonic stages, that imposes itself on the totality of zoological forms. It is a tree, but a tree whose branches are of unequal lengths. A simple form runs a short course; a higher form will have a long, and hence more elaborate, embryological road.

If, as we have seen, the study of fossil crocodiles led to a "transformist" vision that at some points recalled Lamarck, it will not do at all to consider

Geoffroy Saint-Hilaire a direct precursor of Darwin. In fact, for the latter, the temporal dynamic is fundamental; within a given species, all the small differences between individuals will or will not provide selective advantages, determining future development. Thus the direction of evolution is formed at each instant, through the confrontation of the organism with its environment, without the end being fixed in advance. In contrast, for Geoffroy Saint-Hilaire, all the ends are already determined; they all exist potentially, atemporally, in the form of branches of that immanent tree that sets out the totality of possible transformations. This is basically a static, geometric, vision of organisms, close to the teachings of Haüy.

These details are important, for a number of writers have relied on certain writings of Geoffroy Saint-Hilaire—and of Serres—to find a proof of evolutionism. Let us remember that later on the latter will be opposed to evolution in the Darwinian sense of that term. At the same time we can understand better Geoffroy Saint-Hilaire's reticence with respect to Lamarck: the idea of evolution is not necessarily wholly contained in the idea of recapitulation.

We give below the complete text of *Principes de philosophie zoologique* (*Principles of Zoological Philosophy*), published by Geoffroy Saint-Hilaire in April 1830. On first reading it is possible to omit the discussion on the hyoid (second part of the chapter "On the Theory of Analogues," session of February 22; "Second Argument" of Baron Cuvier, session of March 22; and "On the Hyoid Bones," session of March 29).

PRINCIPLES OF ZOOLOGICAL PHILOSOPHY

DISCUSSED IN MARCH 1830
IN THE ROYAL ACADEMY OF SCIENCES

BY M. GEOFFROY SAINT-HILAIRE

1830

PRELIMINARY DISCOURSE

ON THE THEORY OF ANALOGUES

To explain how it became the subject of a discussion at the Royal Academy of Sciences, and to specify the exact point of the controversy

For some time studies of organization have been deeply troubled by a malaise that disturbed the way they were conducted; they had gained more in extent than in correctness. Thus a revision of the past became necessary: this crisis was inevitable; that is to say that a serious controversy was bound to break out. That moment has come.

Every renewal of ideas is opposed for an extended time in its upward march by long days of a transitory state: minds are then caught in a moment of hesitation, even of suffering, which makes them linger for the most part in the traditions of the past, but this also becomes a critical moment for the innovators. This indifference, perhaps also some considerations of rivalry, increase their scientific faith and devotion, and excite them to redouble their efforts. From there, from these lively impressions to a declared hostility, there is but a step. If it is taken, the two camps are formed; a violent collision is imminent.

That is what the action, the inevitable influence, of the time on certain ideas, recently produced and, relative to the study of comparative anatomy, has now led to, has caused to break out in the course of March 1830; the daily newspapers and the medical journals have taken account of this scientific event. Thus the press has brought it to the attention of the public that the very lively debates between Baron Cuvier and me have been reverberating in the Academy of Sciences. The great fame of this society, the importance of the subject, and the presence of a very large audience have lent solemnity to our dispute and are the reason for whatever interest has been accorded to it.

It is in these circumstances that I propose to give to the public the speeches the Academy has heard read and to set out the development of the rival ideas in the order of their production. But first I must specify their object.

A first address, which was the object of a very lively repartee, stated a single fact; it was, and would be in the whole course of our discussion, only a question of giving a resolution to the following propositions:

Should we, must we, religiously preserve an ancient method for the determination of organs, in acknowledgment of its ancient and useful services, when, although it has borne excellent fruits, it is now insufficient in extremely complicated cases? Or rather, to satisfy new needs, must we prefer another method that provides that determination with more certainty and more expedition, when the latter is recognized as better suited to that office, as has been proved, since it has already triumphed over what had until then been considered insuperable difficulties?

To be satisfied with this form of exposition would be like trying to come by surprise upon a favorable decision. I do not desire that favorable opinion; on the contrary, I expect it only from a perfect conviction; and for that purpose I wish to demonstrate clearly in what the procedures of the two methods consist, and to show what advantages they definitely possess. A single example will suffice.

The first goal that the two methods propose in equal measure is to know what organs in animals correspond to the organs previously known and studied since ancient times in man. Neither the point of departure nor that of arrival give rise to any uncertainty. All the parts of the human body are known, and it is to discover as well the analogous parts of the bodies of animals, to see them in their reciprocal concordance, that all the researches of comparative anatomy are directed. As many as are found to be similar, so many are the relations whose establishment forms the high points of transcendental anatomy.

Now the two methods have both been applied and have encountered one another in the consideration either of the hand or the foot, the last portion of the anterior extremity. But how are they taken up? That is the point I intend to examine; for if I have been understood on this occasion, I shall invoke the adage: *ab uno disce omnes* (from one case learn all!).

The old method followed step by step what it called the degradation of forms, starting with man, that is to say, with the organization it considered the most perfect. At each moment of its investigation it is at something *approximately similar*, from which it descends to each perceptible difference. It undertakes to recognize these differences; it has no other concerns, no other subject of study. This hand of the orangutan is approximately like that of man, but it differs by a shorter thumb and longer fingers. Following this same kind of reasoning we arrive at the hand of the spider monkey, which is defective in quite a different way, for in one of the species of this genus, there is no longer any thumb, and in another, to take its place, there is a very short tubercle. If we move on to other monkeys, the silky marmosets, the wistitis, for example, the five fingers are observed; the *little short of* still continues; but at the moment of investigating the difference, we come to perceive that it is no longer a hand, in the sense that the internal finger is no longer opposable in its ability to bend toward the other fingers. Those, like the internal finger, are equally minute: they close together, they are equipped with long, hooked, sharp nails; from then on their forms and their functions are profoundly changed; for this is no longer a true hand, but a claw. Wistitis climb up the tree trunks by using their nails. Thus it is by a different mechanism that this little family manages, like all the monkeys, to live in the woods and also to reach the top of the trees.

We pass to the bears; the same reasoning is again invoked. Their paw is again *approximately the monkey's hand*, but with another appearance; the differences here are more pronounced; for what is found to be observed and what has to be described is a *paw*, so that it is named for its condition of nonresemblance, that is to say, a foot with digital parts, short and gathered together, the nails pressed against one another, robust and terminating in a point.

I omit several intermediates to arrive at the otter. Here we observe a new circumstance; the fingers of this mammal are united by large membranes. Thus this *approximately the same thing* has strangely changed forms; and, since it furnishes the animal powerful means of swimming, it is now called a *fin*.

The method goes no further: it finishes with unguiculate mammals, also called fissipede mammals. Now let me remark at this moment that, first, the method is neither logical nor philosophical. What has been proposed that one obtain through it is a tableau of cases of diversity that can serve for the distinction of entities. However, see what has been done by means of a supposition that can be admitted, strictly speaking, in an extended application, but entails a contradiction in the way in which it is expressed. At each instance one is forced to invoke a semiresemblance, a presentiment of relations not justified by an attentive preparatory investigation: a vague idea of analogy is the link to which these observations of different cases are attached. Is it in fact logical and philosophical to behave in this way, to reason from resemblance to difference, without having first explained clearly so much that is *approximately similar*?

Second: This same method offends yet more by its insufficiency. You have stopped at the fissipede mammals; you cannot pursue your comparisons further; and they ought to be extended to the feet of ruminants and horses. But there the differences appear to you too great: as if the method were frightened to make a judgment in this case, it remains silent. That was a guiding thread; it is broken, it will no longer give directions. To avoid this difficulty, we change the system: we pursue our study of these cases of diversity, making use of this language: "Why should nature always act uniformly? What necessity could have constrained her to employ only the same pieces and to employ them always? By what arbitrary rule would this have been imposed on her?" This *approximately a hand*, this part so named in man, can no longer be understood through the same comparisons when it happens, as in the ruminants and in the horse, to be added to the leg itself. But these are not relations that are of concern in this case; it is only different facts that are being investigated. Is there exaggeration in the metamorphosis of the foot of ruminants? So much the better. The description, the only thing we want to provide, will be all the easier to produce, and will show more striking traits. This is even a sort of good fortune for this order of investigation: for its practitioners have sided with the belief in a different plan of animal composition. New names find their way into the descriptions: those of *hoofs*, *shanks*, *ergots*, and so on: this helps to establish admirably that "*Nature does not allow the imposition of any arbitrary rule*."* They proceed to abandon the field of relative differences when the relations are masked: if it would take painful investigations to unmask them, they are satisfied with the observed differences. But to neglect some common points is to admit that the differences are complete and absolute. Nevertheless, who would dare to announce that there are differences presenting this character?

Let us oppose to the proceedings of which we have just been giving an account the conduct that the theory of analogues prescribes for arriving at a severe and philosophical determination of the same organs. First it must give itself a clear and well-circumscribed subject: that is the only means it has to escape the seductive influence of forms and functions, an influence that tends to introduce a number of circumstances where only one fact, which it is a question of examining, must be admitted. Then one is no longer forced to drag oneself along link by link and to invoke those *approximate similarities* where there are no true resemblances. Thus we begin by seeking *the subject* that furnishes its general condition independently of all accessory dispositions, an isolated object that the principle of connections illumines by its torch, and which invariably, despite its possible modifications, retains the fact of its primitive essence, its philosophical character of a uniform composition.

*Cuvier, lecture at the Academy, April 5, the text transcribed in the pamphlet of the *Journal des débats*, dated April 6, 1830.

This offers no difficulty. In all the vertebrates the anterior extremity is composed of four segments: the shoulder, the arm, the forearm, and a terminal segment that forms the hand in man, the claw in the cat, a wing in the bat, and so on. Without pausing for considerations of form and of function, which are entirely secondary conditions for the last section of the anterior member, I see this section insofar as it exists: it is that in the abstract and that alone that I first consider. It will not escape me in this condition, for I am observing it by bringing to bear on it an inflexible aid, the investigative eye of the principle of connections. A barrier is set us by that fixed given: where the third section, that is, the forearm, finishes, the fourth, or the terminal segment of the front member, begins.

With this anatomical element thus isolated, thus disengaged from considerations of forms and usages—considerations nevertheless important if they arise at their level of research—with this element alone, I compare one and the same fact in the whole animal series. I do not stop after the fissipedes; I pass, without the slightest difficulty, to the consideration of the foot of camels, horses, cattle. Everywhere I go on to consider this same anatomical element in birds, in reptiles, in fishes, indeed, in all creatures.

Since I have not disposed of my working hours outside my usual occupations, I am not in the position of conducting myself with respect to Nature, if I happen not to understand her, with some appearance of generosity, in not wanting to refuse her the right and the power to act as she pleases. I have acted up to now in a different manner in order to show myself more reliably her devoted interpreter. In such a case, I scorn the feeble lights of my reason; I take care not to ascribe to God any intention: I remain where it seems to me that an *ordinary* naturalist ought to remain.* I limit myself to the duty of the strictest observation of the facts; I claim only the role of historian *of what there is.* And to explain myself in this matter, I did not wait for this recent argumentation, which is only a repetition of an older exposition. I gave it in a *Fragment sur les existences du monde physique* (*Fragment on the Existents of the Physical World*), which appeared in the modern Encyclopedia (see volume 17, under the heading *Nature.*)

So far, however, by taking the terminal section of the anterior limb as subject of comparative study, I have as yet satisfied only one condition. I must also attend to the vessels that arrive from the forearm to that organ; they have produced it in the first place and continue to nourish it. We understand how the

**For us ordinary naturalists:* familiar expressions of M. Cuvier in the Academy of Sciences; repeated several times, they had the expected effect, but perhaps beyond what was explicitly intended.

principle of connections limits its extent: one of the organs is the generator of the other.

So here is another matter with which the theory of analogues is concerned, or with which it advises us to be concerned. Before embarking on research into the differences, it will have gone through a large proportion of the facts in order to appreciate them in the relations they have in common; it will have noticed in what family, or even in what species, the largest number of materials are found, and in what way they are disposed in relation to one another in virtue of their connection; and it is evident from all these facts that the new method of determination proceeds on organs that are known to be exactly comparable.

With these precautions taken, watch the zootomist engage in investigations of the most dissimilar cases; how he walks with certainty through each of them! How he knows better and more completely their respective value! For, going from one species to another, he appeals each time to all the materials, and takes into account the differences, the absence or atrophy of some and the hypertrophy of others. As a consequence, he is prepared for the singularity of that thumb in the spider monkeys, which is entirely lacking in one species, and which, in another, still exists in the form of a rudimentary tubercle. Thus without astonishment the zootomist runs through all the metamorphoses of the organ he is considering; far from stopping at the foot of the camel or the horse, he could, if necessary, compare it directly with the human hand; for that is a given that can serve as a rule. Everything that follows the third section of the anterior foot forms an ensemble of parts that are related to one another, just as much in the horse as in man.

In this way the precautions taken in the first phase of the investigation, to prevent our missing the real relations, are of benefit to the second phase, which must initiate the study of facts that are dissimilar. Thus knowing in the first place what the relations are, is to prepare oneself to recognize more easily later on, to distinguish more easily to what extent there are differences in a given organ, whether in one species or in another.

This amounts to saying that the old method neglected taking all precautions, and that the new one exhausts them all; that the old method took its point of departure a priori, and that the new one has confidence in its way of proceeding only after it has evaluated it by a posteriori investigations; finally, that the old method believes that the fourth section of the anterior limb is comparable in the fissipedes, *before study* of some elements that exhibit this effect; and that the new method, but only *after study*, after it has tested some determining elements, applies itself without uneasiness to all the characteristic distinctions to be acquired.

Behold: henceforth nothing any longer implies a contradiction: for if, resorting to the procedures of the new method, you wish to give a brief and precise expression for your observations establishing each difference and changes of function, for example, this is henceforth without the slightest difficulty. And in fact you can set up in advance an organ that has a special name, that is an identical entity, inalterable in this point, and independently of all ulterior considerations. This given, does it suit you to enumerate all the changes of function that might have been noticed, and that are no longer anything but special facts relative to the organ chosen as example? You can express yourself with clarity, and in this way: the last segment of the anterior extremity is used in most mammals in a diversity of ways, becoming the paw of the dog, the claw of the cat, the hand of the ape, a wing in the bat, a flipper in the seal, and finally a part of the leg in the ruminants.

I am not presenting all this as new; I have already used this example several times. The debate did not take notice of it; perhaps it had not been forgotten. But if it was not necessary to draw attention to it, that is one more reason for me to return to it. Likewise, I do not hear it said that by means of hesitant steps, and precisely because people were already guided by the new principles, they had not, on their side, arrived with the old method at the same conclusions as the new method with respect to what is strictly speaking the mammalian foot. I do not wish to contest a fact that, on the contrary, I had prayed for. My demonstration was not possible, and is not complete, until I should have chosen, and did in fact choose, my example, to compare the procedures of the two methods in a completed work, in studies followed through to an equal extent and claimed reciprocally by the two schools.

Now the liveliest reproaches have broken out. The new method of determination and the theory of analogues that assists it by its inspiration have done nothing for such a question: it is guaranteed in advance that they will be impotent in such and such a case, contradicted in such another. But are these reproaches in truth legitimate? As to this new method, I present it as an instrument of research: I recommend its use only under that title. And it is actually a true instrument of discovery, if it is always applied with attention to the intimate association of its particular rules.* Finally, they say, it would not be called upon to give such a solution, to procure such a different approach. That is possible.

*The theory of analogues, the principle of connections, the elective affinities of organic elements, and the balancement of organs: See, for the development of these ideas, the preliminary discourse of my *Anatomical Philosophy*, vol. 2.

But further, if that argument was to mean anything, it would be necessary that the Aristotelian principles, to which they recur with so much display, should have done better. Now it is 2,200 years since they were promulgated, since they achieved for zoology the definitively essential foundations that have lasted from that moment (debate of February 22). However, what have they truly, on their own account, introduced into science? With them all the analogies, hidden under the veil of great metamorphoses, were not even guessed to be probable. With them, I may go on to say, in the matter of analogies, it was necessary to stick to the coincidence of three givens that are always encountered in the species participating in natural families: *the anatomical element, the form, and the function*. When this threefold relation was encountered, as between man and ape, the organic structure was considered analogous: the human eye in its essential conditions could be studied in the ape's eye, and vice versa. However, to arrive at this result, was it necessary to refer to doctrine, to ascend to *those essential foundations of zoology that Aristotle, their creator, had established for ever and ever?* No, that is my belief, no. Folk common sense had already given Aristotle that instinctive truth, to his century, to the centuries before him that had preceded the time of mature reflections and scientific studies. Folk common sense does that on its own: today, it does it in countries that are not yet civilized, and it will always do it, since the evidence carries in itself a principle of manifestation fit to strike all minds equally.

If you weigh all these reasons, it will be evident to you that the Aristotelian principles, in leading us to the presentiment of certain analogies, were never of assistance as parts of a scientific method: for it is not on reflection, but in an instinctive manner, that the facts resulting from striking analogies are admitted as soon as they are noticed.

Do analogies exist, on the contrary, of a sort such that, while not revealing themselves easily to the eyes of the body, they can nevertheless manifest themselves to the eyes of the mind, and so the Aristotelian principles are insufficient for this case? The old method stops in its applications, just at that moment when it would have to become doctrinal, where it would have to become an Ariadne's thread, to make us appreciate the most hidden relations, all the common points of general facts, the most important points of the sciences.

I have again heard this reproach: "But this new method, so highly recommended, has been only rarely employed." I gladly agree; to begin with, it is not of an ancient date; and then, to cite an example, while I employ it to reveal the various metamorphoses that the facts of monstrosity introduce in the normal arrangement of the organs of a given species, I leave in abeyance all the perfections otherwise possible and desirable. To this, which I do not deny, I reply that I cannot do better. I add: if this method were constantly applied from now on by the agreement of all zootomists, in two centuries it would no doubt have been entirely adequate.

That is what this argument seems to ignore, what it leaves aside, for it has not paid attention to what might have modified its lively attack. But on the contrary, it believes, or appears to believe, that I have presented the theory of analogues as sanctifying the principle of the invariable conservation of all materials. Without producing any justification, it hastens to point out some differences in the number of pieces, when in most cases this is only a product of age, the result of the association of several pieces by suture. To produce a greater effect, all that is observed in the case of that allegation is collected: the proofs abound; one is drowned in the details. No doubt a useless effort, for the theory of analogues accepts all the variable numbers that observation provides; it claims only to study information.

I now summarize the preceding in these terms. The question is not, and cannot be, raised, whether or not I have successfully brought to the theory of analogues all its fruits: that was not in the beginning, and is not actually, the subject of our controversy. The point under discussion is to know if it is mistakenly or for good reasons that I have recommended a method for the determination of organs, and if the latter is preferable to the method that has long been usual.

I have come to grips with these two methods in a well known example: let the judgment be made. If it is objected that in the example in question the old method would have closely followed the new one and that it would have arrived at almost the same results, there is nothing to be inferred from this against the practical utility of the new one, since it is not by the latter alone that the most difficult problems can be resolved, the most peculiar metamorphoses resolved, or all the extraordinary variations comprehended that have been ascribed to the supposition of several plans of animal composition.

But instead of replying to me categorically on this point, my opponents have preferred to divide their attack, to multiply the details, to argue on the basis of the accidents of numerous modifications of bodies, to profess sincerity in enumerating facts attesting to the diversity of animal organization. They wanted to make themselves formidable in order to impose silence, powerful so as to arrive with the advantages of an elevated position, head of a school in order to overwhelm with authority. That is what inspired an interrogation from on high—conduct founded on a clever calculation, on the idea that on my side the arms would not be equal, first, because reprisals are foreign to my character, and second, because there is no way to reply adequately to so many interrogations repeated blow by blow. The Academy listened; it is I, public professor at Paris for the past thirty-seven years,* whom they were not afraid to interrogate on facts, on matters for a first-year course.

In these circumstances, I thought I should put a stop to the debates in the Academy. The presence of a large audience evoked too strongly the desire for a triumph and substituted for an interest in the subject matter an interest too personal to each of us. Therefore I announced to the Academy that I would no longer abuse its patience in listening to us and that henceforth I would print my replies. At the same time the following, first formulated as a prospectus, was distributed to all the members of the Academy.

I pledge myself to a series of publications, with confidence in the magnitude of the results to be obtained. Could it be true in these times of great enlightenment that one should have to take credit for believing in the coordination and concatenation of observations in natural history? To describe animals in isolation, to understand them more or less happily in works of classification: is that enough to do if one dreams of participating in the movement that at present absorbs people's minds? To stick to isolated observable facts, not to want to compare them except in the circle of separate groups or small families, is to forgo the lofty revelations that a more general and more philosophical study of the constitution of organs can induce. After one animal has been described, that is to recommence for a second, then for a third, in other words, as many times as there are distinct animals. For other naturalists, other destinies! They make useful abridgements, and their knowledge is that much more profound if they embrace organization in its highest correlations. For in that case, if account has been taken of all the possible developments, as much of those of a single species traversing the ages of life as of those of the whole zoological series rising

*Three years younger than Baron Cuvier, I nevertheless preceded him by eighteen months in my career of instruction. This circumstance, my position at the Royal Garden, brought us into contact, and led to our relations.

These relations began for us at our entry into social life: they soon became intimate. Then what cordiality, what concern, what mutual devotion there was between us! At present should disagreements about the facts of science, however grave they may be, prevail over the sweetness of those memories? Our first studies in natural history, even some discoveries: we made them together; we behaved with the impetus of the most perfect friendship, to the point where we observed, we meditated, we wrote reciprocally, the one for the other. The anthologies of the time include writings jointly published by M. Cuvier and me.

by degrees to the greatest organic complexity, we arrive at a simple fact, which is at the same time the most general condition of organization. Every organ is brought to the unity of its essence and of its capacity to incorporate certain elements. A simple organ, grafted on another of the same order, introduces the first facts of complication. That several others then come, at their precise moment, and by the paths of succession and of generation, to surround that core: that augments the sum of the first facts without altering the character of their simplicity. But further, it is the same course of development that is pursued in the same circle, satisfying its original tendency. For there is only one mode of formation to engender organic facts, whether its action, stopping early, produces the simplest animals, or whether that action, persevering to the limit of all its possible capacity, leads to the greatest complication of organs. In fact, there would be here no question of miracles, but of the action of time, of progress in the relation of less to more.

For this order of considerations, there are no longer any different animals. One fact alone governs them; it is as one single being that it appears. It is, it resides in Animality, an abstract being that is perceptible by our senses under different shapes. These forms do indeed vary, as is ordained by the conditions of special affinity of the ambient molecules that are incorporated with it. To the infinity of these influences, ceaselessly modifying profoundly the contours as well as points on the surface, there correspond an infinity of distinct arrangements from which arise the varied and innumerable forms spread throughout the universe. Thus all these diversities are limited to certain structures, according to the character of the exciting factors, and according to the way that the elements are displaced and reengaged. But further, these facts of diversity are reproduced necessarily, as if each were restrained and enclosed in a web that it can neither pierce through nor extend beyond.

Such is the ocean of actions, or perturbations, and of resistances in which the powers of animal organization are exercised. The bodies, the elements, their movement, the actual and future arrangement of all things: that is the work of God, those are the gifts he has given for all time.

Nature is the law he has given to the world.*

*Profound thought of the poem on astronomy, posthumous work of M. Daru. "A poem," said N. Lamartine, taking his place in the French Academy, "a poem that was published only yesterday, and which promises to illuminate his tomb with the latest, but the most brilliant, ray of his fame."

This way of understanding nature, of considering it as the glorious manifestation of the creative power and of finding in that immense spectacle of created things motives of wonder, or gratitude, and of love, which constitute the relations and the duties of humanity with respect to the master and the supreme legislator of worlds is, I believe, no less respectful than the form that was admitted in the lecture at the Academy on April 5. I could count on arguments from naturalist to naturalist; the argument became theological;* the desired effect was produced. I shall abstain from relating here the judgment of the public in this matter.

And in fact, among naturalists the word *Nature* is susceptible of only one interpretation: like all natural scientists, they find it, they believe it is given, in the meaning of that phrase: *God is the author and the master of Nature.* It is indeed the case that nature extends universally to all created things.

How is it possible after that to reject this clear and precise meaning, to give it, in the same writing, another sense, to make *Nature* play as well the role of an intelligent being who does nothing in vain, who acts by the shortest means, who never exceeds them and always does everything for the best?

This double meaning is without doubt a resource in a debate; but, for my part, I claim my right, rejecting every application that might illegitimately be made of that extension, recalling and admitting only the signification admitted in natural history.

This is also what was proposed in that other objection, on March 22. "Let us conclude that your supposed identities, that your supposed analogues, *if they contained the least reality,* would reduce *Nature* to a sort of slavery, in which its author is happily far from being bound: we would no longer understand anything about beings, neither in themselves, nor in their relations to one another. The world is an indecipherable enigma."

The following passage is the source, and contains the development, of that thought:

Truly I call Nature the law of the Omnipotent,
Of the supreme father, which from the first origin of the world
He imposed on all things, and decreed and held it
Inviolably, while the ages of the world shall remain.

—Marcel Palingen, *Zodiaque de la vie,* vol. 2

*I know very well, said Baron Cuvier, in his memoir of April 5, I know that for certain minds there lies behind this theory of analogues, if only confusedly, another older theory, long ago refuted, but which some Germans have reproduced for the benefit of a pantheistic system, called *philosophy of nature.*

If it contained the least reality. That is to say, if there were truth in the assertion of that proposition, you would nonetheless reject it! Would it be the case that a fact of natural history does not always constrain the naturalist? Well then! Abandoning ourselves to our own judgment, we could *prefer the best to what is the case.* To congratulate oneself on *Nature*'s having escaped a sort of slavery, that is to let it be understood that the speculations of our weak reason could have some weight, could count as a corrective in the arrangements of the universe, however admirable.

I understand quite otherwise the duties of the naturalist: if he takes seriously everything there is, if he seeks knowledge through observation and if he speaks plainly without claptrap, he is confined to the role of a simple historian of facts, a role from which he is forever forbidden to depart.

You are offended by considerations of utility in favor of the novelty of certain analogies. That is to beg the question. These analogies are or are not the appropriate generalized expression of particular observations; that is the only point which, under the name of naturalists, we are called on to judge. If they are true, even if they should be difficult to grasp, we owe them our assent; if they are false, if they should be of such a nature as to facilitate the first steps of youth, it is fitting to reject them. The majesty of the sciences resides wholly in the respect for truth; and it is evading it, I believe, to argue with reasonings like this.

"It is doubtless more comfortable for a student of natural history to believe that everything is one, that everything is analogous, and that through one entity one can understand all the others: as it is more comfortable for a student of medicine to believe that all diseases compose only one or two" (debate of March 22).*

What students need, as do those who are professionally learned, is to be right. The whole worth of the sciences is there: all sound philosophy rests on that axiom.

Constantly pursued and long matured investigations on the analogies of beings do not tend to make *of the world an indecipherable enigma*!

In the answers by which I shall reply to the arguments that have been opposed to me, I shall concern myself definitively only with what concerns everyone, with science. Never cleverness, always straightforwardness, the awareness of facts, care in their narration, a perfect conviction in their arrangement,

*For my part, I pledge medical students to hold to the instruction they now receive; for if they had to fall back on the nosology of Sauvages, they could not be up to the thousands of diseases distinguished by that practitioner.

sustained work: that is what there will be, what will be found, I hope, in this first publication and in those that follow.

May I, now that I have arrived at the conclusion of this work, have finally acquired the right to add an ordinary signature, this last word at least expressing the sentiments that animate me and sustain me in my research: *utilitati*, to utility.

Paris, The Royal Garden, April 15, 1830.

N.B. I give a date for this first article, the day it was sent to the printer. Although intended to illuminate a point of the controversy, as *preliminary discourse*, it summarizes some parts of it.

ON THE NECESSITY FOR PRINTED WRITINGS

To replace verbal communications by this mode of publication in controversial questions

It was urgent: it was necessary to bring these successive pleadings to a stop as soon as possible, and I had recourse to the printing of a prospectus in which I announce that from then on I would treat the subjects in dispute only by using the voice of the printed word. My prospectus, distributed on April 5, 1830, to all the members of the Royal Academy of Sciences, expressed my thoughts in the following terms, the text of which I reproduce here.

To my regret, I find myself engaged in a polemic with Baron Cuvier on fundamental points in the science of organization: on his side, my learned colleague testifies that he is as tired and distressed as I am. In these circumstances, friends of both us, and colleagues, speak of intervening: they think that it is time to stop this struggle of opinions that keep colliding in successive pleadings. In fact it could become even more lively, and finally compromise such long-standing relations of friendship, founded on reciprocal service and esteem.

Some people have imagined, and say, that our dissension bears chiefly on obscurity and on a confusion of badly defined terms that the slightest concessions would easily cause to disappear. They are wrong about that: there is a great, essential, truly fundamental fact at the bottom of things, which gives a soul to natural history and calls the generalizations of that science to become henceforth the first of philosophies.

Always describing, without making the descriptions issue in some practical utility: that is a past for which it is the propensity of present-day minds to guarantee the future. Special considerations abound to overflowing: let us show gratitude to those who prepared the way, but further let us rejoice in so many accumulated treasures. Advances in public opinion demand that we use the facts today, chiefly to know them in their interrelations. Let us really practice science.

Thus I will have to persevere in the defense of the ideas of mine that are under attack, of a doctrine that a deep conviction tells me it is necessary to produce, even at the moment; but, what seems to me in every respect preferable, I can do it by more inoffensive means. For to continue our *impassioned* struggle would be to lead rather to the disparagement of science than to the triumph of truth.

Through preference for having recourse to the path of publicity through the printed word, our discussion will be debated before the people who are the most enlightened on the subject: thus I address myself to the only judges who can know with full competence the points at present under litigation. In this manner I can only respectfully await an ultimate decision from that high tribunal.

N.B. When, two weeks ago, I wrote that last paragraph of my prospectus, I was not unaware of the fact that in Germany and in Edinburgh there was consideration of new theories of the philosophical resemblance of beings. There we have been overtaken, there they pursue without respite, with conviction, with perfect confidence in their success, what we in France are trying with so much reserve, doubtless with too much timidity. There is a general movement, a decided engagement of minds in those doctrines that are finally included. And truly, I would be unjust not to recognize it, it is the same in France, where some celebrated curricula conform to it. Such is the instruction in anatomy at Montpellier (under Professor Dubreuil), and that of the natural history of animals at Strasbourg (where Duvernoy is professor), and so on.

But better yet: while these questions were being agitated with great eclat in Paris, and in the Academy of Sciences, where it was being recommended with such vehemence that the torrent be resisted, that we should defend ourselves against new ideas, it was at that same moment, in the Academy of Sciences, that that severe lesson had to be received, that the dike they had tried to set up was decidedly impotent. Zoological anatomy, now strengthened by other principles, could not be turned back to the traditions of the past.

And in fact, works conceived and carried out in the spirit of the new school, maturely reflected on, and above all foreign to the present controversy—for they had begun some months earlier—such works, I say, have just been communicated to the Academy. They have been addressed to it, not in any way as linked even indirectly to our debates, but as brought to light necessarily by the development of human faculties, hence as brought to light by the advancement of science. Now there is in this present conjuncture a fact doubtless curious enough, so that it should not be surprising that I remark on it, and that I make known its principal circumstance.

Dr. Milne Edwards comes (April 1830) to present to the Royal Academy of Sciences an extensive work, "On the Organization of the Mouth of Suctorial Crustaceans." This memoir, communicated to some friends six months earlier, was thus not in the least destined in principle by its author to take its place and color in the present controversy; but it is tied to it by its form, its expressions, and its general tendency. "We know," the author says, "two principal groups of crustaceans, crustaceans who live a wandering life, and who have their mouths armed with strong, cutting organs of mastication, and crustaceans who live as parasites, whose mouths are destined to give passage to liquids. Thus this is a structure apparently entirely different: for the observing eye, this is the spectacle of two plans of animal composition. Here the mouth is surrounded at the same time by jaws and by cutting mandibles; there, it is considerably elongated, and, becoming a tube, it is transformed into a sucker." The conclusion of the memoir is that *the organic composition described has always remained analogical. The same constituent elements are discovered in the one and in the other case; there is a remarkable tendency to uniformity of composition.*

M. Savigny has presented a similar work and has given the same demonstration with respect to insects compared to one another.

The comparisons of M. Edwards's work are followed through perfectly, the relations are deduced from it with certainty, and his demonstration is complete.

How will the argument directed against the analogies of organization, persisting to the conclusions of his thesis, be able to accept these results, if, as I cannot doubt, they appear to him to be certain? I believe I hear this reply:

> It is in the embranchement of articulated animals, and better yet, it is in a given class in that embranchement, that of the crustaceans, that these mouths, difficult to recognize in their excessive metamorphosis, have been studied. From there they could, by an effort of sagacity, be reduced to a common conformation. But what is strictly possible between beings of a given embranchement presents an incommensurable difficulty if the comparison is attempted between animals belonging to two very different embranchements.

This reminds me of the pains that a military man, in the high ranks before 1789, gave himself in my presence in 1795 to demonstrate to some friends that the army of Sambre and Meuse could try in vain to cross the Rhine opposite Düsseldorf. "What obstacles there were! The width of the river, the difficulties of the location, the fortifications of the city, the defense batteries, etc. Who would dare to enter into battle? These would no doubt be the undisciplined masses gathered on the left back, bands directed by unknowns who had come out of the crowd, people from nowhere, who were called *Jourdan, Kleber, Bernadotte, Championnet, etc.* Let some one try, in good time, in the face of the enemy, the passage of some small rivers in the interior of country, that would be to accomplish remarkable feats of arms. But to throw oneself at great rivers like the Rhine, that is temerity, that is folly." While such speeches were being made in Paris, the Rhine was crossed, and the city of Düsseldorf was occupied by the French. It had been said to simple people, and they believed it; but trained minds denied that it could be done.

It seems that, when it comes to investigations of the analogy of organs, enterprises are sanctioned that are calculated on a scale corresponding to the passage of small rivers, but that then any other enterprise, equivalent to the military crossing of a river as great as the Rhine, is forbidden, not only as excessively dangerous, but as downright impossible.

So great is the gap between its extreme terms, so imposing is the hiatus between the families at the base and at the summit of the zoological scale!

REPORT PRESENTED TO THE ROYAL ACADEMY OF SCIENCES ON THE ORGANIZATION OF MOLLUSKS

(SESSION OF FEBRUARY 15, 1830)

1. *On the report, as having given rise to the controversy.*

Did I really open the hostilities? And to what extent? This point of fact seems to me to have excited some curiosity; thus an explication is desired. I shall give it by presenting in published form the writing by which the susceptibility of Baron Cuvier was offended, and which was followed on his part, on February 15, by an improvisation as ardent as it was bitter.

Two anatomists, MM. Laurencet and Meyranx, had applied six months earlier to give a lecture to the Academy. In order that I might try to give them a favorable opportunity, they asked me to take account of their subject and of the interest of their memoir; but when they were tired of waiting, they finally asked the president of the Academy to have their manuscript examined. M. Latreille and I were given this assignment. The very next day, February 9, the authors saw one of their reviewers. They were delighted to learn that I had completed so lengthy a task as I had been engaged in, and that, finding myself free to go on to another, I had leisure to take account of their research. On the following days, we observed, we dissected together; and, so that I need not return to it later when I would have other duties, I wrote the report at once, the ideas for which I had just been gathering. Hence, if this report was written in the interval between one meeting and the next, there was no haste as far as I was concerned, but only convenience relative to the hours I could dedicate to this work.

In order to explain how the investigations of MM. Laurencet and Meyranx occurred at that particular time, according to the needs of the day, I explained historically what had been established previously, and with success, as to the facts in question. Where I thought I was putting into place the elements of a eulogy, Baron Cuvier saw an allusion and the intention to wound him. No less surprised than distressed by his comment, I protested that that had been far from my thought, and at that moment I put all the sincerity of which I am capable into asserting this once more. In a friendly spirit, I offered my learned colleague to suppress either the whole report or whatever parts he chose. He accepted my offer for a page that I at once removed, and M. Cuvier was the first to demand a vote on the report.

Here is the report, as the Academy adopted it. Some warmth may be perceptible, thanks to the enthusiasm of my conviction, but nowhere, I flatter myself, nowhere can any envious hostility be found.

2. *Parts of the Report Offered for Consideration and Adopted by the Academy.*

Last Monday you received from MM. Laurencet and Meyranx a first memoir on the mollusks under the title: *Quelques considérations sur l'organisation des mollusques* (*Some Considerations on the Organization of the Mollusks*). M. Latreille and I, to whom you assigned this task, are now going to give you an account of it. *Some considerations,* a vague title, but probably without color through excess of modesty, since, promising only new efforts following older investigations, this title contrasts all the more with the results that the authors flatter themselves they have obtained. In effect, if their avowed claims are well grounded, what these authors have found is order where their predecessors, by their own avowal, had seen only confusion. This is the key to an organization described, but not yet understood in its composition. It is the philosophical resemblance of organisms reduced to a common measure, which until now the masters of science had only signaled as out of order, as unintelligible: which, in this case, amounted to saying that the law of these paradoxical existences rested on a constant statement of as yet inexplicable monstrosities.

However, if such were the difficulties of the subject, how did MM. Laurencet and Meyranx, after so many vain attempts, so many fruitless meditations, decide even to raise such questions? The discoveries, whatever they may be, need preliminary explanations if they are to be understood and appreciated; so it becomes necessary to give all the pathways, to state all the intermediate ideas with which the mind comes to be concerned. This means, on the part of the discoverer, to take the trouble to let every one know of his new procedures—that is, by a kind of replacement of the parts, to address himself to the sagacity and to the enlightenment of the true judge in all things, the Public.

The importance of the question being treated, and even more the duty that you have imposed on us, prescribes to your reviewers that they act in the same way under these circumstances.

What precedents were in fact favorable to the authors, so that we could accord them our confidence on the point that they announced they had examined? In what did the first works consist that they had already published? In simple essays, it must be said, but essays that, in truth, bear on the most important systems of organization, like the brain and the spinal cord. As to those essays, however, if they held the attention of the public for a moment, were they not occupied much more by the singularity that characterizes them than by their true originality?

But nevertheless that was not their only point of departure. We have seen in the hands of MM. Laurencet and Meyranx a great number of drawings already lithographed and ready for publication in the near future—drawings that represent new facts in anatomy. These gentlemen estimate them at three thousand figures, and this enumeration is doubtless not exaggerated. But these drawings bear on difficulties of science, for they give the zootomy of many animals of the center and of the lower levels of the zoological series, such as salamanders, fishes, crustaceans, insects, but above all the anatomy of certain mollusks.

Let another reflection dispose us just as favorably! In all the works of the intellect, there is a propitious hour for them to be conceived, developed, and matured. Before our young authors, activity was only by instinct. The talent of the zootomist, whatever his powers, had before his eyes only bizarre forms: he was obsessed by them and such causes of inspiration necessarily captured him.

Thus, for example, one of the cephalopods, a cuttlefish, had its mantle brightly colored, and its other parts of the body pale and as if deprived of insolation. For an uninstructed observer, there was the back and here was the belly. This was perhaps to have judged correctly as to the situation of the animal moving in its external world. However, who could have guaranteed this determination with regard to the superposition and to the relation of the organic parts of the being in itself? But formerly this was not even the object of a question.

It became so later in particular for MM. Laurencet and Meyranx. They put their trust in a guide that was available in science, in a method of determination that offers its principles to produce the desirable inspirations and revelations, which promises the authority of its past successes in order to offer good supervision in whatever judgments arise—so that, thanks to the process prescribed by the new method, the investigations are instantaneously scientific.

Formerly one animal was looked at and anatomized, then another, then a third, etc., and the only a priori that served the mind was the idea of seeking, observing, comparing. It was fortunate then if some common points that had been clearly obtained emerged from these efforts. You took a chance of rising, randomly, to the character of a general proposition. But at present, with the help of the new method of determination, those important results of science arrive at the same time that the facts generating them are being investigated.

Thus it was formerly good luck to encounter procedures with a broader validity, while today we arrive without hesitation at the very foundation of our questions.

Through the preceding reflections we have not intended to depreciate the merit of the work we were charged to examine, but by recalling how the authors got help from what is at present going on in science, to give rise in people's minds to a presumption in their favor.

So much the more do we owe them that support, since one of us, M. Latreille, in 1823, had also attempted, for his part, to lift the veil that had until then hidden the relations between certain mollusks and some animals of the higher classes. In fact, M. Latreille placed a work that appears to be unknown to MM. Laurencet and Meyranx in the first volume of the Memoirs of the Society of Natural History of Paris. This work of March 14, 1823, has as its title: *De l'organisation extérieure des céphalopodes comparée avec celle de divers poissons* (*On the External Organization of Cephalopods Compared with That of Various Fishes*). Four views of this relation are contained in well-summarized propositions.

That said, we pass to the considerations contained in the memoir of MM. Laurencet and Meyranx. Those able anatomists, confident that they are sufficiently prepared and informed by the research attested to by the numerous figures of which we have spoken, and which they believe already form a certain presentation of their views, have proposed as general facts the following propositions:

1. Every mollusk presents, in an envelope more or less deprived of solid parts and of systems of sensibility, a *vegetative* system recalling that of one or several higher animals.
2. The viscera of which these organs are composed are placed in *the same connections* as in the higher animals, and their functions are carried out by *a mechanism and by motor organs that are similar.*
3. The connections that seem to be inverted are so only in appearance. The key to finding the unvarying persistence here is furnished by the consideration that the mollusks' trunk while still retaining a longitudinal situation, is found, on the contrary, folded toward the middle, and that in return the two parts, welded to one another, are reversed at one time on what is called the ventral face and at another time on the face called dorsal.
4. The orifices in question are revealed at the outside by the respective position of the orifices.

5. Finally, in the case of the parts that are resistant and are caught in the skin, those earthen masses are again comparable to certain bony parts in the vertebrates.

Wishing to give the justification of these theoretical views, MM. Laurencet and Meyranx applied them to the order of cephalopods, and even, to make their thought clearer, to one species in particular, to the cuttlefish, *Sepia officinalis*.

Everybody knows the cuttlefish. Imitating in this respect MM. Laurencet and Meyranx, we must avoid describing them in terms borrowed from the organization of other families, if those terms give rise to false connotations. A great flattened pouch, circular at the bottom, presenting a large opening with cut-off edges, and composed of two surfaces, the one brightly colored and slightly convex, and the other white and flat, forms the principal part of that animal. A rounded mass, which begins with a narrow neck and is terminated by eight fleshy tentacles, comes out of the mouth of the large pouch, as if at the bottom of a funnel. The buccal orifice, armed with a beak like the beak of a parrot, is in the center of these appendages; then ahead and on the sides are two large eyes. The rounded part coming out of the constriction of the pouch is determined by all the experts to form the head of the animal. And given that the means of locomotion, consisting chiefly in the tentacles of which we have just been speaking, are distributed around the mouth, and consequently toward the terminal part of the head, the cuttlefish and its analogues that have the same form of locomotion and are characterized by that peculiarity are called *cephalopods* or *feet-in-the head*—a denomination that is due to Baron Cuvier, and which he imposed on that family when he was laying the first foundations of his zoological fame, that is to say, when he elicited from a shapeless chaos the classifications of the mollusks, so rightly admired and at once adopted by the learned world of Europe.

However, what characterizes the cuttlefish, as its specialty and singularity, is this point of fact: that it is a soft animal, or, in other words, that it is an entity belonging to this degree of organic formation, that an arrest in development had confined it to this first level of vital power. All these circumstances have as their effect that its secretions produce no saline molecules, or at least only a few, to become, in consequence of the arrangements of animal organization, so many bony molecules: at least only a few, we said, for we know the bone of the cuttlefish. This much given, the folding asserted by our authors can be considered possible.

But would this folding, as to its proper disposition, be happily explained by a thought of our authors stated as follows?

> The first idea that emerges from the bizarre and abnormal situation of the cephalopods, which have their cloaca placed on the nape of the neck, is that these animals walk and swim by presenting their vertex either to the ground or to the bottom of the water, and that all their organs that present analogies with those of higher animals are arranged on a plan that we believe can be rendered by this very simple formula: *Let us imagine a vertebrate walking on its head; this would be exactly the position of one of those tumblers who turn their shoulders and their heads backward so as to walk on their hands and feet;* for then, in this reversal, the extremity of the animal's pelvis would find itself attached to the posterior part of the neck.

Let us take this only for an image producing a first and crude explanation, for otherwise this comparison would push us, through an entirely natural inference, to false analogies. Thus, for example, because of an entirely similar function, we would be led to consider the tentacles of the cuttlefish and the limbs of higher animals to be reduced to the same interrelations, to the same essence of organization, since, according to the determination given by one of us, M. Latreille, in his memoir, referred to above, these tentacles represent only the barbs surrounding the mouth of the catfish. Hence in the cuttlefish this would only be that same apparatus, brought to the maximum of its possible development, and acquiring, through the benefit of its greater volume, more numerous and more important functions.

A point at which our authors have wisely halted is to have preferred to the consideration of forms that are transitory from one animal to another, and bad counselors for philosophical comparisons, the indications given by the principle of connections. And in fact it is in the spirit of that philosophy that MM. Laurencet and Meyranx have paid great attention to the situation of the diaphragm. That is what they call an extended muscular lamina, quadrangular, placed parallel to the mantle, attached at the sides, that occupies the central position among the viscera. Following these authors, the viscera skirt round the posterior edge of the diaphragm and are thus spread out on the two surfaces which they call, by reason of that circumstance, the one, the gastric face, and the other, the branchial face. They add: "The folds of this central muscle are promptly recognized, bordering the esophagus, perhaps even the psoatic muscles, which would be found in two strong muscular cords at the base of the large pouch, where they occupy a lateral and posterior position."

The study of this diaphragm would already be very useful, but one could, and doubtless should, make more of it, for if the determination of this muscle were really made with accuracy, on its own it would first have to become a point of departure for all other desirable determinations. Employing the usual thread, the principle of connections—so fortunate a guide!—, it would be necessary to go on from there to recognize and group methodically around the diaphragm all the other organs that are attached to it by their superposition and the agreement of their functions.

Would our authors not have already done enough for the establishment of their thesis? Perhaps. But at least let us be grateful to them for having embarked on so competent an undertaking: their work reduces the jaws to their natural position—it would have been said that they were positioned upside down. They see in the cartilaginous ring of the neck the elements of a hyoid, and those of a pelvis in certain stylets, also cartilaginous, that border the base of the funnel.

We will not follow MM. Laurencet and Meyranx further in their efforts of determination. In so serious a matter it is appropriate for us to remain reserved, and to insist, in a first report, only on the more or less probable degree of the correctness of their views.

And indeed, how can we not believe in some similarity of organization when we encounter, enclosed in the same integuments, organs as elevated in their structure as are two venous hearts and one arterial one, a perfectly regular collection of gills, medullary material concentrating chiefly in front of the neck, a very extensive liver, perhaps a spleen, if Meckel's conjecture is admitted, but more probably, on the word of our authors, a system of vessels secreting urine, the latter consisting, they continue, in a spongy tissue served by an excretory channel, prolonged and open in the cloaca? Further we find, equally associated and located together, a whole intestinal apparatus, a beak constructed like that of the parrots, the esophagus, all the generative organs, closely imitating those of fishes: can it be said of so many things that this is a totality interconnected altogether differently, combined altogether differently? To prove this proposition, that is, to demonstrate that there is here only one fact of a great, surprising anomaly, there would be more to be done than to maintain the contrary thesis. For we would have to admit that these organs, which can exist only if engendered by one another, and because of the mutual fitness of nervous and circulatory actions, would give up belonging together, or being all in agreement with one another. But such an hypothesis is not at all admissible, since, if there is no harmony among the organs, life ceases: then there is no animal, no longer any animal. But if on the contrary life persists, it is the case that all those organs have remained in their habitual and inevitable relations, that they work together as they ordinarily do. And then, moving from consequence to consequence, the fact is that they are linked by the same order of formation, subject to the same rule, and that, like everything that is animal composition, they would not know how to escape the consequences of the universal law of nature, *the unity of organic composition*.

MM. Laurencet and Meyranx knew how to appreciate the need of science, since they tried to diminish the hiatus apparent between cephalopods and higher animals. No doubt they did not hope to arrive at first at a completely satisfactory result, but we owe them at least the justice of saying that they are happily trying to open the road, and that they have even traversed some of its paths. Their leading idea is ingenious, and if we agree to consider their work only as interesting studies to contribute to the natural history of mollusks, under that heading their memoir appears to us worthy of being included in the compilation of works by nonmembers. We have the honor to make this proposal to the Academy.

Signed: LATREILLE, GEOFFROY SAINT-HILAIRE, *authors of the report*

3. *Part of the report that was withdrawn, but is here reproduced exactly.*

The altercation on the subject of the report had appeared to the Academy, to those present, and to M. Cuvier himself, to have been put to rest by my friendly explanation, by my easy concession, and the agreed-on suppression of one part of my writing. Later advice changed these first arrangements.

M. Cuvier, by his argumentation on February 22, changed these first impressions. "M. Geoffroy Saint-Hilaire," he said, "has seized avidly on the views of MM. Laurencet and Meyranx; he has announced that they *refute completely* what I had said on the distance that separates the mollusks and the vertebrates, etc."

The elements of such a bitter sentiment were doubtless not to be found in my report as printed above; nor will they be found in the piece of that same report that I had read, and of which I had been eager to allow the suppression. However, it matters to me that people should be plainly convinced of this, and so I am obliged to return to that portion, which I had kept, and which I here give literally, as follows.

The part of the report that had been suppressed was placed after the words "*the unity of organic composition*."

> However, at the beginning of the nineteenth century it was possible, and doubtless necessary, to produce an entirely contrary philosophy. In a fragment

rich in facts, powerful and bursting with knowledge and sagacity, all the cases of difference characteristic of the cephalopods are listed that are considered to lead to the consequence that in respect to them there is no resemblance whatsoever, no analogy of arrangement in the repetition of the same organs. This document concludes as follows: *In a word, we see here, whatever Bonnet and his disciples may have said, Nature passing from one plan to another, producing a whole, leaving a manifest hiatus between her productions. The cephalopods are not the transition to anything: they did not result from the development of other animals, and their own development has produced nothing superior to them.*

Let no one misunderstand the meaning of these words, chiefly the motive that makes us repeat this quotation. Science was then already what it was proper for it to be in every epoch of its cultivation: philosophical, large, progressive. But it was no longer pointing to the only goal of a zoology to be founded or at least to be improved—and that is precisely because from 1795 to 1800 it had so fortunately attained that goal that, always faithful to the character of its essence, to its needs of extending and acquiring improvements, it was now pursuing another goal, which placed it beyond the first one. In fact, its object today, its greatest needs now, in view of the preoccupation of present-day minds, consist in the knowledge of the philosophical resemblance of beings.

Thus, zoology will first have demanded the greatest rigor in its classifications: for their sake it had to begin by stressing dissimilar facts with a sure hand. In fact, to try to introduce greater precision in characteristic distinctions was to undertake to present with more brilliance and more success *the table of the animal kingdom*, everything greater and more imposing for philosophy that the enumeration and registration of the productions of nature have produced. We shall add that it is not before this Academy that it is necessary to recall that such an enterprise has at one and the same time become a *French work* and one of the greatest achievements of our epoch. But it is nevertheless true that the beginning of the nineteenth century will remain remarkable through this tendency in its studies, through the preference that was then given to the investigation of differences.

Now I urge the reader to take the trouble to weigh the value of these statements, which I have repeated without modifying them, and to make a judgment.

That phrase, transcribed from an earlier writing, where *the opinions of Bonnet and his disciples* are recalled with disfavor, caused some irritation to be experienced. I need not have reproduced it; it has even been as good as decided that I had no right to do so. This is what made people say that I had expressed myself without assuming the moderate tone that the sciences demand, and failing in the courtesy that belongs to every well-brought-up person.

I also had to reply to other complaints. In speaking of the *French work*, would I really have been exceeding the conventions of behavior by an excessive measure in my praise?

FIRST ARGUMENTATION, OR CONSIDERATIONS ON THE MOLLUSKS

BY BARON CUVIER

(SESSION OF FEBRUARY 22, 1830)

The publication of my replies would fail to achieve their goal if I did not keep my readers informed about the observations and doctrines to which they were replying. A happy circumstance offers me the means of doing this. M. ***, a young disciple of M. Cuvier whose devotion to his master knows no bounds, and whom the administration of the *Journal des débats* (*Journal of Debates*) had made its collaborator for the section on the sciences, devoted to the addresses of my learned colleague the major portion of the issue of his journal the very day after the academic sessions. If it is not the whole, it is the chief and the most important part of the memoirs that is there transcribed word for word. Thus I believe I can do no better than to refer to these extended extracts; and as I am doing today, I shall be able to draw further on the same source for the further addresses of M. Cuvier.

Extract from the Journal of Debates

We have spoken (so the journalist begins) only a word on the discussion that arose in the last session between MM. Cuvier and Geoffroy a propos of a report made by the latter on a memoir of two young naturalists who had presented some new ideas on the organization of the cephalopods. These singular animals, placed by M. Cuvier at the head of the genus of mollusks, were compared to mammals by MM. Meyranx and Laurencet, by means of a fiction that seemed very ingenious to the gentleman reporting on their work: they had supposed that they (the cephalopods) were folded into two by being bent over toward the front, and that rearranging them in thought was sufficient to place their organs in the same situation where we find them in the mammals. M. Cuvier did not allow the report of his learned colleague to pass without appealing to the opinion that he had expressed and maintained in his works, and which finds itself contradicted by this new work; but it was impossible to present all the sufficient explanations in a few instants. It is to clarify completely this interesting point in the history of mollusks that M. Cuvier read in today's session a memoir that is distinguished by a method and a clarity both of which are perfect, and by that charm of style that characterizes all the works of the author. We believe we will give pleasure to our readers by providing an extended analysis of these interesting considerations.

Text of the first argumentation, given in extracts

The mollusks in general, but more particularly the cephalopods, have an organization that is richer than that of other invertebrates and in which we find more viscera analogous to those of the higher classes. They have a brain, often eyes, which in the cephalopods are even more complicated than in any vertebrate; sometimes ears, salivary glands, multiple stomachs, a very considerable liver, bile, a complete double circulation furnished with auricles and ventricles, in a word, with very vigorous impulsive powers, distinct senses; very complex male and female organs, from which issue eggs in which the fetus and the means of nourishment are disposed as they are in many vertebrates.*

These different facts had already resulted from the observations of Redi, of Swammerdam, of Monro, and of Scarpa, observations that I have extended much farther, supported by numerous preparations, and of which I have availed myself for the past thirty-five years to establish that animals so richly

*Did I go too far in my report? Did I distinguish in the mollusks a superfluity of *analogous viscera*? Was I heard to call such a great number of organs, all declared the same, by the same name? According to the argumentation, the mollusks enjoy an *organization that, by the abundance and diversity of its parts, approaches that of the vertebrates*, and yet in another article the mollusks would be presented *as being the passage to nothing*!

But there is at least a very large hiatus between the mollusks and the fishes! Without the least doubt! In the same way the Seine is narrower at Paris than it is at Rouen. By this I mean that in the latter case this is easily known, while on the other hand, appreciating the interval that separates the mollusks from the fishes is very difficult. It will take the cooperation of several naturalists to succeed in this. Moreover, that is the object of M. de Blainville's research, and it is the same thing that M. Latreille has done in his memoir of 1823. What did these *two young and ingenious observers*, MM. Laurencet and Meyranx, expect from their recent efforts? To collaborate in the same way in that work of naturalists.—G.S.-H.

provided with organs could not continue to be confused, as they had been before me, in a single class with polyps and other zoophytes; but that they should be distinguished in their nature and raised to a higher degree on the scale, an idea that seems to me to have been adopted today in one way or another by the universality of naturalists.

However, I took good care not to say that this organization, approaching, through the abundance and diversity of its parts to that of vertebrates, was composed in the same way, or arranged on the same plan. On the contrary, I always maintained that the plan that is common to that of vertebrates up to a point does not continue in the mollusks; and as to the composition, I have never admitted that it could reasonably be called *one*, even taking it in a single class, or a fortiori in different classes. Quite recently, in the first volume of my *Histoire des poissons (History of Fishes)*, I have again expressed my sentiment on this subject, doubtless in the tone of moderation that the sciences demand and with the courtesy that belongs to every well-brought-up man, but nevertheless in a manner so clear, so positive, that no one could have misunderstood me.

The question, with its proofs, is open to the eyes of all naturalists; it is to them that it belongs to pass a judgment on it, and I would have abstained, as I have been doing for ten years, from addressing the Academy about it, if a circumstance to which that body was a witness had not forced me to renounce a resolution that had been dictated to me by the desire to employ my time more usefully in contributing to the progress of science, and by the persuasion that it is by a deeper appreciation of facts that the truth in natural history is more likely to come to light.

Two young and ingenious observers, examining the way in which the viscera of cephalopods are placed relative to one another, had the thought that one would perhaps find, between these viscera, an arrangement similar to that familiar among the vertebrates, if the cephalopod were represented as a vertebrate whose trunk was folded over on itself toward the front at the level of the navel, so that the pelvis would approach the nape of the neck; and one of our learned colleagues, seizing avidly on this new view, announced that it refutes completely what I had said about the distance that separates the mollusks from the vertebrates. Going even much further than the authors of the memoir, he concluded that zoology has had up to the present no solid base; that it has been but an edifice built on sand, and that its only basis, henceforth indestructible, is a certain principle that he calls *unity of composition*, and of which he is confident that he can make a universal application.

I shall examine this question in its special relation to the mollusks; in a sequence of further memoirs I shall treat it relative to other animals. I hope to do this with the same urbanity that my learned colleague has used toward me; and since the writings he has been directing for ten years against my point of view have never in any way altered the friendship I entertain for him, I hope that he will likewise be among those to whom I shall now successfully defend these ideas. But in every scientific discussion, the first thing to do is to define well the expressions one is employing; without this precaution the mind promptly goes astray. Taking the same words in one sense at one place in the argument and in a different sense in another place, one produces what logicians call syllogisms of four terms, which are the most deceptive of sophisms. Moreover, if, in the exposition of these same arguments, instead of the simple language of words used in their proper meanings, the language rigorously demanded in the sciences, use is made of metaphors and rhetorical figures, the danger is greater yet. You try to get yourself out of an embarrassing situation by a trope, reply to an objection with a paronomasia, and by thus turning aside from the straight path, you are at once caught in a labyrinth without issue. But I beg the Academy's pardon: I see that I am losing myself in the language I am rejecting, and I hasten to return to that in which I shall continue to speak for the rest of this memoir.

Let us begin then by coming to an understanding on those grand words of *unity of composition* and of *unity of plan.*

The *composition* of a thing signifies, at least in ordinary language, the parts of which that thing consists, of which it is composed; and the *plan* signifies the arrangement of those parts among one another.

Thus, to use a trivial example, but one that expresses well the ideas in question, the *composition of a house,** that is, the number of suites and of rooms

*I had used the same comparison in September 1829; this was also in order the better to express my thought. That was when I was writing the article *Nature* for the Modern Encyclopedia, a work to which, as editor, M. Courtin, a former magistrate, devotes his leisure hours. In adhering to the request of that learned jurist that I should write the article *Nature*, I found the occasion naturally appropriate to reply to some critical remarks of another article on *Nature* that M. Cuvier had formerly placed in the *Dictionnaire des sciences naturelles (Dictionary of the Natural Sciences)*, published by *Levrault*. M. Cuvier there addressed to me the following objection:

> These views of *unity* have been repeated thanks to an old error at the heart of pantheism, engendered chiefly by an idea of causality, by the inadmissible supposition that *all beings are created in view of one another;* however, every being is made for itself, has in itself what concerns itself.

that are found there, and its *plan* is the reciprocal disposition of those suites and of those rooms.

If two houses contained a vestibule, an antechamber, a bedroom, a living room, and a dining room, we would say that their *composition was the same;* and if this bedroom, this living room, etc. were all on the same floor, arranged in the same order, so that one would pass from one to the other in the same way, we would say that their *plan was the same.*

But if their order were different—if those rooms were on one level in one of the houses and on different stories in the other—we would say that with a similar composition those houses were built on different plans: thus the *compo-*

To this objection I replied as follows:

But who doubts that? Yes, without doubt, an animal inevitably forms a finished whole, given that the respective position and reciprocal agreement of its parts are the conditions of its anatomical structure, given that the way in which it finds itself established consists of obligatory properties, special as well as harmonious. It is quite simple that its organs, its actions are of this sort.

At present I am searching, but I do so in vain, for whatever connection could have been perceived between these ideas that no one contests, and those declared above to be a false product of the mind, engendered by ideas of causality. From the relations that I perceive among the materials that recur identically to compose animals, from these data that produce a certain resemblance in all entities, as much internally as externally, I arrive at a deduction, at a general idea that includes all these coincidences; and if I embrace and express these under the form and the name of *unity of composition,* I am proposing by that only to translate my thought in a simple and precise language. But further, I am careful not to say what I am ignorant of, that one thing would be made intentionally for the sake of another. In these conclusions I believe myself definitively founded on reason, just as, seeing as a whole the numerous buildings of a great city, and keeping myself to the common points that are the conditions of their existence, I would go on to reflect on the principles of the art of architecture, on the uniformity of structure and of method in such a large number of buildings: one house is not built *in view* of another; but all can be reduced *intellectually* to the unity of composition, each being the product of identical materials, iron, wood, plaster, etc.: and the same for the unity of functions, since the object of all is equally to serve as human habitations.

Every organic composition is the repetition of another, without its being in fact produced by the successive development and transformation of a single nucleus. Thus, it does not occur to anyone to believe that a palace was first a humble cabin, which had then extended to make a house, then a mansion, and then finally a royal edifice.

The science accomplished at one point is composed of generalized facts, and consequently of philosophical relations. And it is such results that are told off as proclaiming more or less probable opinions, that are even condemned for finding themselves too decidedly placed under the reflection of the romantic inspirations of a Talliamed!— G.S.-H.

sition of an animal is determined by the organs it possesses, and its *plan* by the relative position of those organs, or by what our learned colleague calls their *connection*.

But what is the *unity of plan*, and especially the *unity of composition*, which are supposed to serve from now on as a new foundation for zoology? That is what no one has yet told us clearly, and yet it is on that that we need first to fix our ideas.

A person arguing in bad faith would take these words in their natural sense, in the sense they have in French and in every other language; he would claim that they mean that all *animals are composed of the same organs arranged in the same way;* and, starting from there, he would soon have pulverized the alleged principle.

But it is not I who will suppose that even the most vulgar naturalists could have employed those words, *unity of composition*, *unity of plan*, in their ordinary sense, in the sense of *identity*. No one of them would dare to maintain for a moment that a polyp and a man had in this sense *one single composition*, *one single plan*. That leaps to the eye. For the naturalists of whom we are speaking, *unity* does not mean *identity;* it is not taken in its natural usage, but it is given a twisted sense to mean *resemblance*, *analogy*. Thus, when we say that there is *unity of composition* between man and the whale, we do not wish to say that the whale has all the parts of a man, for it lacks thighs, legs, and feet—but only that it has most of those parts. This is an expression of the kind that grammarians call *emphatic;* in this case *unity of composition* means only a *great resemblance in composition*.

In the same way, when they say that there is unity of composition between man and a snake,—the snake, which has no anterior extremities, and whose posterior extremities are reduced to mere vestiges—, they only mean to say that there is a *certain resemblance of composition* between them, but less than between whale and man.

It is evident that it would be a formal contradiction in terms to call *one* or *identity* a composition which, by the admission even of those who employ these words, changes from one genus to another.

What I am saying about composition also applies to plan; we would think we were doing an injury to those naturalists if we claimed that, by the words *unity of plan* they meant anything but *a greater or lesser resemblance of plan*. If not, opening a bird or a fish before them would suffice to refute them in an instant.

But once these extraordinary terms have been thus defined, once they have been stripped of that mysterious cloud in which the vagueness of their meanings or the twisted senses in which they are used envelops them, we arrive at a result that is doubtless quite unexpected, since it is directly contrary to what has been put forward.

For far from furnishing new foundations for zoology, foundations unknown to all the more or less gifted men who have cultivated it up to the present, restrained as they were within the usual limits, on the contrary, they form one of the most essential foundations on which zoology has rested since its origin, one of the principal foundations on which Aristotle, its creator, placed it: a foundation that all the zoologists worthy of the name have sought to enlarge, and to the solidification of which all the efforts of anatomy are devoted.

Thus every day we can find in an animal a part we did not previously know, and which allows us to grasp some analogy between that animal and those of other genera and other classes; there can even be new connections, relations newly perceived. The work that is done with this result merits our praise; it is by such work that zoology broadens its foundations; but take care not to believe that research with these results makes zoology transcend its foundations!

If I had to refer to some work worthy of all our esteem, it would be from that of our learned colleague M. Geoffroy Saint-Hilaire that I would select it. When, for example, in comparing the head of a mammalian fetus to that of a reptile or an oviparous animal, he recognized that relations in the number and arrangement of the parts were apparent that were not observable in heads of adults; when he learned that the bone called quadrate in birds is the analogue of the auricular tympanum of mammalian fetuses, he was making very important discoveries, to which I was the first to do full justice, at the time of the report I had the occasion to make on it at the Academy. Those are further traits that he has added to the resemblances of varying degrees that exist in the composition of different animals; but he has done nothing to add to the ancient and familiar foundations of zoology; he has in no way changed them; he has in no way proved either the unity or the identity of that composition or, finally, anything that could furnish a new principle. Between some closer analogy in certain animals and the generalization contained in the assertion that the composition of all animals is one, the distance is as great—and this is to say it all—as between man and monad.

Thus we all know, and have known for a long time, that the cetaceans have at the sides of the anus two small bones that are what we call the vestiges of their pelvis. Thus there is there, and we have been saying that for centuries, a resemblance, a slight resemblance, of composition; but no argument will persuade us that there is unity of composition in the case, since this vestige of a pelvis carries none of the other bones of the posterior extremity.

In a word, if by unity of composition is to be understood identity, that is to say something contrary to the simplest testimony of the senses; if it is *resemblance*, *analogy*, that is intended, something is being said that is true within certain limits, but as ancient in its principle as zoology itself, and to which the most recent discoveries have done nothing except to add, in certain cases, more or less important traits, without altering anything in its nature.

But in reclaiming for us, for our predecessors, a principle that has in it nothing new, we take good care—and that is how we differ essentially from the naturalists we are combating—we take good care not to regard this as a unique principle; on the contrary, it is only a principle subordinated to another, much higher and much more fruitful, to that of the conditions of existence, to the agreement of the parts, to their coordination for the role that the animal has to play in nature;* that is the true philosophical principle from which flow the possibilities of certain resemblances and the impossibility of certain others; that is the rational principle from which that of the analogies of plan and of composition is to be deduced, and in which, at the same time, it finds those limitations that some wish to ignore.

*I do not know any animal that HAS TO play a role in nature. In my view, that idea is far from forming a principle to be recommended; on the contrary, I see in it a grave error against which I stand up unceasingly with the sense of rendering an important service to philosophy. Let us take care not to explain what exists necessarily after reversing the terms of the argument. In this abuse of final causes, that is to have the cause produced by the effect. Thus, on noticing that a bird transverses regions of the atmosphere, you would conclude that it is given an organization to suit that destiny? You would admire how in fact, in order to weigh less, it has hollow bones and an ample covering of light feathers, how its anterior extremity at a given point is extraordinarily enlarged, etc. I have also read, on the subject of the fish, that because it lives in a medium more resistant than air, its motor forces are calculated to afford it such and such a means of progression; that because it forms part of the embranchement of vertebrates, it has to have an interior skeleton. To argue this way, you would say of a man who uses crutches that he had been destined from the beginning to the misfortune of having one of his legs paralyzed or amputated.

To look at the functions first, and then afterward at the instruments that produce them, is to reverse the order of ideas. For a naturalist who draws conclusions after the facts, every being has left the hands of the creator with the appropriate material conditions: it can act according to the power granted to it: it uses its organs according to its capacity for action.

But this observation would take me too far; I shall come back to it in a moment: I return to my subject.

With all that I have just been saying posited and agreed on—and, I repeat, it has been agreed on and posited since Aristotle, for 2,200 years—naturalists have nothing else to do, and in fact they do nothing else except to examine how far this resemblance extends, in what cases and at what points it stops, and whether there are beings in which it is reduced to so little that we can say it is absolutely finished. This is the object of a special science that is called comparative anatomy, but which is far from being a modern science, for its author is Aristotle.

I shall take the liberty of submitting, from time to time, some chapters of this work to the Academy; but I request their permission to offer only some considerations on the cephalopods, a subject that has most happily been chosen by our learned colleague; for there is no one who can see more clearly what the principles of the discussion contain that is correct, and what they contain that is vague and exaggerated.

Suppose, he has told us, that a vertebrate is folded at the point of the navel, bringing together the two parts of its spinal column like certain tumblers; its head will be close to its feet, and its pelvis behind the nape of its neck. Then all its viscera will be placed relative to one another as they are in the cephalopods, and in the latter, they will be placed as they are in vertebrates that have been folded in this way. That part, which because of its brown color you call the back, will correspond to the anterior half of the abdomen; the base of the pouch will correspond to the umbilical region; what you call the front of the pouch will be the posterior or lower half of the back. That more projecting jaw, which you take to be the lower, will be the upper jaw; everything will come back into place: unity of plan, unity of composition, all that will be demonstrated.

I shall say first that I do not know any naturalist so ignorant as to believe that the back is determined by its darkened color or even by its position at the time of the animal's movements: they all know that the badger has a black belly and a white back; that an infinity of other animals, especially among the insects, are in the same situation; they know that an infinity of fish swim on their sides, or with the back below and the belly above.

But to recognize the back, they have a more reliable criterion: that is the position of the brain.* In all the animals that have one, it is above; and the

*I regret having to reply on the situation of the *brain* of the cephalopods; I suffered much more when, within the Academy, I was called on loudly and arrogantly to explain myself on this point. However, no impediment, no circle of Popilius, troubled my mind:

esophagus and intestinal canal are below. Our learned colleague himself has noticed this in one of his earlier memoirs. For us as for him, that is the true criterion, and not a puerile remark about the color.

Starting from there, I have taken, on the one hand, a vertebrate; I have folded it, as was requested, with the pelvis toward the nape of the neck; I have removed the integument from one side, in order to exhibit plainly its inner parts; on the other hand, I have taken an octopus, I have placed it beside the vertebrate, and I have taken account of the respective situation of their organs.

It is true that in this position the more projecting jaw of the octopus corresponds to the upper jaw of the mammal; but to come firmly to this conclusion, the brain would have to be placed toward the gullet, as it is in the mammal toward the nape of the neck. But it is quite the contrary: the octopus's brain is situated toward the face opposite the gullet.

That is already a terrible presumption against the idea that the gullet is a pelvis folded over to the nape of the neck.

But let us continue. In order that the side on which the gullet rests should be the side of the nape, it would again be necessary that the esophagus pass between this side and the liver, as we see it doing in mammals; but again, it is quite the contrary; it passes on the opposite side, on the side that we call dorsal . . . , etc.

Now I ask: how, with these numerous, these enormous differences, fewer

---

other cares held my attention. I was hesitating to give the correct reply. What confusion, what storms could follow! I stopped at the idea of not wounding an old friend.

And in fact, to say to M. Cuvier that the cephalopods lack a brain, that the demonstration of this fact has recently been given, that science has made new observations on the nervous system of those animals, and that he, classical author on this matter, unfortunately retained false prejudices in favor of his thesis of 1795, which was true in some respects, but also too broadly generalized: that is what I did not feel I had the courage to declare before the large audience who were present at this debate.

In placing, for such good reasons, the mollusks some degrees higher in the zoological ladder, M. Cuvier found himself drawn beyond the facts; one ought not to assign to those animals a place higher than that of the insects. This point is universally accepted in Germany, and the work of M. Serres on the nervous system of the cephalopods puts the decision beyond doubt. With respect to their nervous system, the cephalopods must be ranked below the insects and the crustaceans; for their cephalic ganglia are united in the same way as in the doris, and the progression of the funiculi of their nerves is more or less interrupted. On the whole, says M. Serres in his *Comparative Anatomy of the Brain*, vol. 2, p. 34, the mollusks, as to their degree of composition, are beings that do not transcend the larvae of insects. G.S.-H.

on one side, more on the other, could one say that between the cephalopods and the vertebrates there is *identity of composition, unity of composition,* without diverting the words of the language from their most manifest sense?*

I reduce all these facts to their true expression, in saying that the cephalopods have several organs that they share with the vertebrates, and which in their case exercise similar functions; but that these organs are differently arranged in them, often constructed in a different manner, that they are there accompanied by several other organs that the vertebrates do not have, while the latter also, on their side, have some that the cephalopods lack.

I admit that in saying this I am not saying anything that many others have not said before me. But if I do not have the merit of novelty, I flatter myself that I have at least that of truth and accuracy, and of not muddling the minds of be-

*It is necessary to agree on what the terms signify: let us do what has been so well recommended in the course of the present argument. I admit the facts here stated; but at the same time I deny that they lead to the idea of another kind of animal composition. The mollusks had been placed too high on the zoological scale: but if they are only embryos of those lower degrees, if they are only beings in which fewer organs come into play, it does not follow that their organs lack the relations required by the power of successive generations. The organ A will be in an unusual relation to the organ C if B has not yet been produced, if the arrest of development, striking the latter too early, has prevented its production. That is how there are different dispositions, how there are different constructions for ocular observation.

If they *formed a passage to nothing,* the cephalopods would be an eternal objection to the principle of the *necessary* concatenation of the facts of nature in our zoological series! And now this has been affirmed for the reason that between them and the animals least distant from them there is a larger gap than usual! But is there not something more scientific than such a result of observation taken as an absolute anomaly? There are anomalies, for the philosophical naturalist, only as relative: providing difficulties and attacking established theories, they oblige us to modify them. With that given, what is there that is essential, what is there that is true to be considered in the case of the cephalopods?

Every organic part is the product of two systems, the *blood system* and the *nervous system:* both, in their successive developments, follow one another regularly. There is nothing like this in the cephalopods: the usual state is here defective. The blood system here takes on a more extensive development; the development of the nervous system is less extensive. Since their nutritive and reproductive organs, increased through the hypertrophy of the blood system, had been the subject of the first studies, it was necessary to place them fairly close to the fishes, while quite recently, in view of the atrophy of their nervous system, they have been again placed at a lower level. Today, balancing the strong with the weak, cephalopods and mollusks are thought, as before, to occupy a line parallel to that of the insects. G.S.-H.

ginners with undefined expressions which appear, in the vagueness that surrounds them, to present a deep meaning, but which, if analyzed more closely, are either entirely contrary to the facts or signify only what has been said since time immemorial with more or less detail in its application.

In my further communications, I shall examine several other principles and several other laws put forward by various naturalists; but so that these lectures are not limited to metaphysical questions, I shall take care that they are always attached, like today's address, to some determinations of fact from which science can come to a conclusion more solid than those vain generalities.

EXTEMPORE REPLY*

To the first argument of Baron Cuvier at the same session of February 22

I had considered as entirely exhausted the susceptibility that M. Cuvier had shown in our last meeting. Everyone there, and I myself in particular, had believed M. Cuvier was guided by a concession made with the openness of genuine friendship; unfortunately, that is not the case. The cloud raised between us has not been dissipated. In this there is for me a just subject of affliction and regret. But further, I cannot help feeling a certain satisfaction, when I see my learned colleague at last raising serious questions that, until now, each of us had understood differently and on which it seemed to me useful that we should explain ourselves. I am not at all prepared to treat on the spur of the moment all the questions that have just been raised, and I will confine myself today to presenting briefly some preliminary remarks.

1. I applaud the steps taken by M. Cuvier which lead us back to those brilliant moments of the former Academy of Sciences where all the noble subjects of our knowledge were successively brought forward and illuminated by a profound discussion. It is good, in fact, that we leave the new path we had embarked on and break the bad habit of receiving and of hearing without discussion the memoirs presented or read to the Academy. Thus, in place of those discussions that succeeded one another with distinction, vivacity, and profit for every academician, in place of those debates that were always instructing and sometimes happily inspiring, we have now a kind of session in which every communication is sterile, since every one takes care to conceal his opinions.

The admission to our ordinary meetings of some persons tolerated under the name of auditors has gradually changed the older custom. The number of

*I have found in the medical anthologies and in my recollections the principal parts of the talk I gave after the lively attack that precedes. I could not avoid giving that improvisation here, since my written replies, which follow, refer to it. In addition I find there for my readers the advantage that they will have a better knowledge of the events and accidents of our first engagement: *first,* I say, for there was in my *Report* on the mollusks neither the form nor the tone of a provocation.

auditors has grown continually, and it is now—as it has been for some years—before the public that the ordinary sessions of each Monday are held. Since then there is still more reserve in the communications from member to member; the necessity of writing with a little more solemnity; negligence or timidity as to the bearing of those little facts acquired since yesterday, and in which one sometimes had occasion to perceive the germ of very great discoveries. But today, on the contrary, everyone bringing his memoir appears to communicate it only to fix the date, to depose it in a place of public archives, until the day of its introduction in academic collections.

If I point out these inconveniences, it is not that I am asking for their suppression by violent means or that I desire that we operate by declaring the sessions secret in the future.

No; other times, other customs. Under other aspects the presence of the public has several advantages. The encouragement of research is more direct and attains its goal more promptly; the relations of member to member perhaps gain in seriousness from these new circumstances; and without explaining myself further about this, I remain of the opinion that the advantages greatly outweigh the inconveniences: what is ought to be and will thus be maintained.

But there was, and there is, something better to be done: that is the maintenance of the advantages and the disappearance of the inconveniences. Without being disquieted by the great number of witnesses present, let the academicians take more assurance, and let them do in the presence of a numerous public what they used to do in a small committee in the old Academy, and everything will be for the best. Our customs will adapt themselves more and more.

So, here is the example that M. Cuvier has just given us. For my part, I applaud it and I am doing more than simply saying so, by at once presuming to address to you the present observations.

2. About the basis of the argument I shall not today try for long the patience of the Academy. I see there two distinct things, two questions, one that concerns the two young scholars whom it has seemed to me useful to encourage, and another that concerns me personally.

First: Had MM. Laurencet and Meyranx exceeded by much the proper hour for bringing the mollusks into the general scope of science? Did they in fact, through their new and ingenious idea, understand better than their predecessors the organization of those animals? This concern is their affair, and I leave to them all the responsibility, that is to say, all the duties, the dangers, but also the honor of a reply to be produced. As for me, I have praised them only for having entered courageously on a new path of investigation, of having asked for a deeper comparison of organisms in new relationships. That was their due, and I still congratulate myself on having given it to them well and vividly, for I still believe there is merit in their principal opinion.

Without the least doubt, I have acted with a lively preoccupation of mind, but I do not reproach myself with either too much benevolence or with frivolity. The considerations which, even at present, I cannot abandon are: that large and important organs exist in the mollusks as in the fishes, and that they are given the same name, but further, that this is done with good reason, since the principal organs there take similar forms and fulfill the same functions. Although some information has not yet been obtained by the progress of philosophical studies, the recognized points of resemblance no less remain authenticated relations. Then what to conclude from these relations, of them and from them? The fact is—I venture to say this as a presentiment, to decide entirely a priori—that so many similar organs cannot be found in the mollusks in a sense manifestly contrary to one another and so as to give the spectacle of a different system of animal composition.

I said in my report, and I persevere in saying the same thing, that there is a better chance of our having to admit the supposition that the mollusks are reduced to some degree to the unity of composition, than in favor of the conclusion that we can never succeed in showing this.

Second, the argument attacks directly the basis of my doctrine, the questions of the unity of organic composition. Would this in fact be, as this attack gives one to understand, only one of those false doctrines, unfortunate product of illusory propositions, of alleged philosophical chimeras, such as the abuse in the use of good things has so unhappily and so often brought to birth?* This con-

*I learn that the *Considérations sur les mollusques et en particulier sur les céphalopodes* (*Considerations on the Mollusks and in Particular on the Cephalopods*), that is to say, the whole memoir to which the present article was verbally replying, has been printed the same day, February 22, in the *Revue encyclopédique*, to appear in the issue of April 1830, volume 46. I cannot allow myself to be disturbed by the fact that this publication is appearing without the pleadings I have opposed to it, when I consider that that vast collection has long contained the strongest arguments in favor of my doctrine. Doctor Pariset has given the general principles, vol. 3, p. 32; M. Flourens has also devoted to it an article in vol. 5, p. 219, under the title *Essai sur l'esprit et l'influence de la* ANATOMICAL PHILOSOPHY (*Essay on the Spirit and the Influence of* ANATOMICAL PHILOSOPHY), an article in which its author concludes that "the philosophical course prescribed from now on to the science of comparative anatomy will facilitate its direct and rigorous application, and will ensure that M. G.S.-H. will have acquired all species of perfection, for he will have gen-

cerns me alone, and I shall take care of it personally. It is known that this is the happy, or unhappy, dream of my scientific life. All my research has converged on this, the work of forty years undertaken with courage and pursued with per-

eralized and popularized it." I shall also quote a third article from the *Revue encyclopédique*, published in the issue of February 1823, vol. 16; it is by M. Frédéric Cuvier. The last paragraph of that article appears to have been written under a wholly prophetic inspiration. My difficult circumstances today will ensure my pardon for quoting it verbatim.

> A work (*Philosophie anatomique*, vol. 2) filled with so many facts, so many new insights, in which one is carried so far away from the beaten path, cannot fail to excite great interest and to give rise to numerous and lively discussions. Power in the sciences is no longer to be achieved without combats to sustain and rivals to conquer. Whoever wishes to enter the lists of knowledge also needs force and perseverance; but at least in these combats for the truth all efforts are useful, all tend to make the truth appear livelier and more resplendent; also, without having equal rights, they do have incontestable rights to esteem and recognition. In debates of this sort, it is time that clarifies and posterity that judges; and if it is sometimes permitted to anticipate posterity in applauding, that is when authors like M. Geoffroy offer eminent service to the sciences, to *Utility*. *Utilitati* is in fact the epigraph of the work of which we have just been giving a rapid analysis. F. CUVIER, *Revue*, etc.

I shall conclude this note by remarking that the zeal of my friends has not in the least cooled in these distressing times of our dissensions, since, while showing himself full of regard for a very exalted scientific position, and above all of that just deference due to a colleague in whose courses one has lectured, M. Flourens has not recoiled from the difficulty of speaking for his part on the respiration of fishes, the facts of which are precisely and at the present moment under discussion before the Royal Academy of Sciences. So, at least, I gather from the memoir that M. Flourens read last April 12, and from the following extract that I borrow from the account which, on the next day, the thirteenth, the *National* gave of the academic lectures of the previous day.

> After the reading of this memoir, M. Geoffroy Saint-Hilaire asked to speak. One would have thought this would be to remark that his colleague had just translated and done for functions what he had discussed and established in the next to last session on the subject of the conformation of fishes: the unity of functions in fact follows from the memoir of M. Flourens, as the unity of organic composition had been the subject of M. Geoffroy's memoir on the theory of analogues. But the honorable member confined himself to recommending the tuna, *Scomber thynnus*, for the examination of his colleague, as affording new and more powerful proofs in support of his theory. The red flesh and the great vitality of the tuna are simultaneous with the excessive volume of the gills of that fish.

severance. That is what it would be regrettable to have done fruitlessly, but I am not yet reduced to that point. The words I have just heard have in no way shattered my intimate conviction; that is all I can allow myself to say at this moment. I shall defend what is proper to my doctrine otherwise than by this allegation, and I shall do it in a memoir that I flatter myself I shall offer next Monday.

ON THE THEORY OF ANALOGUES

To establish its novelty as doctrine and its utility as an instrument of research

(SESSION OF MARCH 1, 1830)

I am about to reply to the argument directed in the last session against my writings and especially against certain rules I have laid down in natural history.

If one intends to direct one's mind to forgetting allusions meant to injure, one must firmly desire that perfect independence that leaves everything to the care of the things themselves. I flatter myself that I will have that strength of character.

I am even acting without painful effort. The points to be resolved are questions vital to philosophy, and it will easily be understood that they alone should occupy my mind, and that I can only be sensible of their influence on the moral improvement of society.

I have no aspiration to a success that would depend on a talent for fine speaking. I shall employ neither art nor rhetorical precautions in my presentation: I wish to remain with the truth, as much for myself as for my listener. I will also proceed in such a way that the simplest common sense can follow me without difficulty, and come to apprehend, without effort and without delay, the slightest error or the most trivial defect of judgment that might escape me.

For that result, I need only recount what my successive preoccupations have been, to show myself acting under the development of my ideas, and to group together the themes that have led me to imagine the principles of a doctrine that is most certainly my very own, and which I have made known under the name of *theory of analogues*.

When I entered the professoriat, in 1793, there had never been any instruction in zoology in Paris. Commissioned to create everything, I acquired the first elements of the natural history of animals, arranging and classifying the collections confided to my care. However, in order to keep definitively fixed on the best system of classification I could follow, I had first to take account of the value of *characters*, that is to say, to investigate, by long and painful efforts, what characters could offer me that was constant and useful in differences appropriate to serve in the distinction of beings. Now, in every period that I spent daily in the cabinets of the Royal Garden, I received an impression that, constantly repeating itself identically, led me to this intellectual view: that is, that so many animals, which I held to be different and was treating as distinct by giving them specific names, nevertheless differed only in some slight attributes, more or less modifying a structure that was generally and evidently the same. There was in fact only a slight modification, as soon as I saw clearly that the point differentiated did not affect what could be called the essential condition of the parts; it affected only their respective dimensions. Thus, when it came to neighboring animals, each of their organic materials reappeared in its totality. In order that there should be a diversity of species, the smallest variation in the proportional volume of the associated and constituent materials was sufficient: the weakest alteration in dimensions, which changed in no way the essential relationships. And it is to the point that even a slight nuance in color sometimes suffices for the distinction of two beings, like what we see with the stone marten and the marten, for example: two species that are never confused, but which nevertheless differ only in the shade of their throat, which is bathed with yellow in the marten and entirely white in the stone marten.

How often I have taken account of the value of these ideas in thus studying the whole of the collections of the Royal Garden! When I managed to place myself at a certain distance, I grasped a general effect in which all the differences of minor importance disappeared. Looking at the display cases of ornithology, I saw on the shelves only the repetition, multiplied a great number of times, of the type *bird:* that is to say, that I was distinguishing only the general traits, to wit, the head, the neck, the trunk, the tail, the wings, the feet. In all the individuals there were wings as integument; in all, a horny beak surrounding the jaws: everything repeated exactly, and, what is more, which existed in places that were respectively the same.

This same experience, undertaken with regard to the mammals, demanded that I place myself at a greater distance if they were all to be included in the same considerations; and in the same way, by a wholly natural progression, it was necessary to remove myself much further from the subjects to be observed, if I proposed to comprehend under the same aspect, and with the same goal in my research, animals characterized by more numerous and more considerable differences, such, for example, as could be offered by the simultaneous observation of a mammal, a bird, a lizard, a tortoise, or a frog. For in this same case, the quantity of their differences, although giving rise to an impression of larger intervals, or gaps, between these same animals, were no less based on a quantity of difference much inferior to the sum of the relationships by means of which these animals belong, and are arranged, in the same class, and form part of the same group, of the *embranchement of vertebrates.*

Those were my first impressions as a zoologist. Dissections undertaken under the influence of these impressions were in accord with them; all the internal organs were in a perfect relation to those of the exterior of the being. A flow of arterial blood arrives at a given point to lead to its definitive formation every part of that exterior; but to provide for such a regular distribution, there have to be on the inside a quantity of complex contrivances, where the observer might think there was something inextricable, but where everything has its reason, where all things are perfectly coordinated. There is an identical arrangement of analogous systems, so that the zootomist arrives at the same point in his impressions and beliefs as the zoologist, and that it is definitively a well-established fact of natural philosophy that animals are decidedly the product of an identical system of composition, the assemblage of organized parts that repeat themselves uniformly.

Isn't this what you mean, says the argument that has been opposing me: "It is a truth that obtains within certain limits, but in its principle as old as zoology itself and to which the most recent discoveries have done nothing except in certain cases to add more or less important traits without altering its nature"?

I shall examine separately this precise and special point in the discussion. But I will not follow the argument on a distinction that it has attributed to me and which I never made, when it seeks to have us agree on the meaning of those big words, *unity of composition* and *unity of plan.** All that was previously ac-

*If I judge of it by the communications, whether written or verbal, in the improvised summaries, it would seem that the reply I made about *those big words* was considered very weak, or else the objections of which they were the object was considered very powerful. It was repeated with confidence. They ransacked the dictionaries: the word *composition* has a different value from that of *plan,* and vice versa. It is announced that I have admitted on the one hand *unity of composition* and on the other *unity of plan:* so, add up the sum of these two unities, and there is that whole philosophy reversed. The system of nature is no longer *philosophical unity;* nothing would be true except a system of a *plurality of things.*

True, I had seen in the discussion on this subject only a purely grammatical debate about words. However, would I, in my expressions, really have enunciated the distinction that was attributed to me? That would have been contrary to my intention. Let me state my thought more exactly: the *composition* of the parts, without being the same thing as their *relation,* includes or rather calls forth the latter, as its necessary consequence. My

knowledged as true, that is to say: the fact that animals are the product of a single system of composition, I called *unity of organic composition*. To be more precise, I should have said: *unity of system in the composition and arrangement of organic parts*. But I wanted a name, and I could be satisfied only with the contraction of that phrase, thus following the usage that makes us say, for example, *criminal tribunal* instead of the phrase *tribunal instituted to judge criminal cases*.

There would doubtless be a good deal to say in favor of the expression *unity of organic composition*, even to justify that of *unity*. I limit myself to recalling the example given by Leibniz, that definition of the universe, *unity in variety*. But let us leave words and concern ourselves with things.

I return to the observation that was made to me: "Is it really," they said, "only of those resemblances, of those analogies, that you want to speak?" Yes, I reply, that alone constitutes the doctrine of the analogy of organs, and I hasten to add that, at the beginning of my career, I did not believe there was anything there for which I had personally to gain acceptance. Like M. Cuvier, who at the present time makes this the subject of a new consideration, I had myself formerly admitted "that, far from furnishing new foundations, foundations unknown to all the more or less intelligent men who have cultivated them up to the present, these ideas of the relation between beings, confined within reasonable limits, formed, on the contrary, one of the most essential foundations on which zoology has rested since its origin, one of the principles on which Aristotle, its creator, situated it—a foundation that *all zoologists worthy of the name* have sought to enlarge, and to the consolidation of which all the efforts of anatomy are devoted."

I shared these intuitions early on; and in order to be even more deeply imbued with them, I did not content myself with Aristotle's account. At first I had never failed to cite Aristotle in my works, as the first source of the analogy of organization; but I wanted to receive more advanced instruction from the facts themselves. I applied myself for a long time to their appreciation; I asked my study for details, my understanding for direct observation, elements of a

principle of connections serves me as a compass and guards me against error in the investigation of identical materials. Thus, according to the nature of each organ, the same subject of observation recurs in every animal, and provides its condition of an element, of the unity of composition, and subsequently it is for a necessary reason that it is placed in such and such relations, that is to say, under the governance of constant connections with respect to the materials that are adjacent to it. I see here neither confusion in the terms nor obscurity in the thought.

deeper conviction. Then the facts acquired and reflected on in this spirit, compared with one another, and revealing to me relations where until then only absolute differences had been perceived, finally brought to birth in me that awareness of the relation among things, as clear as it is profound, with which at present I am so deeply imbued.

And how would I not have walked in the footsteps of so many able masters? Aristotle's doctrines had also become the reflected insight of all the superior minds who had concerned themselves with this matter since his time.

Thus in 1555 Belon looked at the skeletons of a man and a bird, and he tried to show the correspondence of the same parts, if not yet with similar names, at least with common letters.

Bacon, who remarks in his *Novum Organum* on all the practical benefit that can be derived from an examination of the diversity of beings, nevertheless believes that one could penetrate better into the depths of things if one sought the reason for animal composition in the facts of analogies and similarities. He also recommends, as what ought to become the most useful quality in philosophy, a certain active sagacity that makes us capable of investigating and of grasping physical conformities.

In one of the preliminary discourses of my *Anatomical Philosophy,* I have already called to mind a sudden inspiration of Newton's. His genius, meditating on the relations of the uniformity of the planetary masses, was suddenly struck by the idea that the same views could be equally applicable to animals: "And yes, as well," Newton writes at the close of *Opticks,* "yes, without the least doubt, animal organization is submitted to the same kind of uniformity. *In corporibus animalium, in omnibus fere, similiter posita omnia* (In the bodies of animals, in absolutely all of them, everything can be assumed in the same way)."

Finally, is it not on the idea that the beings of a given group are connected by the most intimate relations and are composed of analogous organs that the scaffolding of the methods of natural history rests?

Such were the ideas at the origin of zoology; I received or conceived of them early on. But they struck me with total conviction especially toward the middle of my career. That is also what Baron Cuvier really thinks about this according to the terms of his memoir, quoted above. After that, one would have to conclude that we are close to agreeing on the foundation of things.

But no; I have to grant to M. Cuvier that there is disagreement between us, and I even believe that this disagreement is greater than M. Cuvier himself suspects. Let us try to make its exact nature clearly explicit.

First, M. Cuvier admits philosophical resemblances, analogies of organs, only within narrow limits: where he thinks to restrict these considerations, I believe it necessary to extend them to a much greater number of animals. But that itself can be explained through a double cause: (1) by what fits better, judging from the fact in itself; (2) by what the susceptibility of the natural qualities of minds decides about it, since some prefer to apply themselves to the superficial extent of things, and others seek to know them in depth.

In the second place, M. Cuvier thinks that science was created 2,200 years ago. Supposedly, Aristotle then placed it *"on a foundation that it is today given to zoologists worthy of the name only to enlarge."* In conceding to me that I had myself made a useful comparison, a very real discovery, when I reduced the heads of mammals in the fetal state to that of adult oviparous animals; and principally when I discovered the analogy of the quadrate bone of birds with the auricular drum of mammals, M. Cuvier declares that I had only added to the ancient and familiar foundations of zoology, that I had not changed them in the least, that I had in no way proved either the unity or the identity of any composition of organs, or in fact anything that could furnish a new principle. Between some further analogy in certain animals and the generalization of the assertion that the composition of all animals is one, the distance is as great—and that is to say it all—as between man and monad.

For my part, I believe all these assertions are far-fetched, even erroneous. But I do not want to rise against them except with the feeling of a benevolent esteem, when I reflect that toward the middle of my career they had even constituted the basis of my own opinions. From 1812 to 1817 I had known all the difficulties of my subject; I had during that lapse of time the weakness of considering them insurmountable. Thus, in a case on which I shall explain myself later, I thought I was stopped by an obstacle that my imagination enlarged to a kind of desperation, where I saw a sort of Chinese wall. I am sorry to say that I took the position of doing or writing no more about these questions.

However, thanks to persistent investigations of fishes, and struck by some luminous insights, I emerged from those difficulties: I came to grasp some precise clarifications, whose happy influences finally produced a collection of ideas, a collection that has today become my theory of analogues.

And indeed, nothing rigorous had yet followed from what it was customary to call, and what I myself had long entitled, the Aristotelian doctrine. But to stick to the truth, I say this again. I need not, and I cannot, concede that the writings of Aristotle are a source of doctrines on the analogy of organization. Thus it would be to stray still further from the truth of the facts to ascribe to that great man everything that the research undertaken in our time has that is severe and truly scientific, and actually to attribute to him doctrines, or a theory, in virtue of whose discovery philosophical anatomy is entitled to take its place among the exact sciences. The fact is that in bowing to the evidence of some observations, Aristotle had presented a priori what I have concluded with certitude and have now deduced a posteriori from the comparative study of the facts themselves. What he had left for the instruction of the ages to come consists in some complex and confused ideas, some true and some false. The former excited the sympathy of superior minds, and at the same time the latter found echoes that led to a predilection for the study of differences.

Now is that a complete doctrine, worthy of interest as such, when it is only a mixture of givens that are mutually exclusive and which, in consequence, vitiate at their very root the proper principle of the analogies of organization? It is clear that this same confusion exists, as much as it did with Aristotle, in the argument that has been opposed to me. For organs were used there for all they offer to the observer; in them, in them first, as to their conditions as organic elements, their forms and their functions were at the same time seen as inseparable.

Certainly Aristotle's idea, as it had been understood during the centuries that had elapsed since his time, was bound to lack lucidity. It was obvious, and people had held to it since its origin. Immediately put into practice, no better beacon was known, no more perfect instrument, more useful in works of zootomy. But it was a different matter in a large enough number of cases. Open the works of veterinarians and of ichthyologists, and you will see there that naturalists use a special language, as if they believed in a special anatomy, as if they understood that they were speaking of organs that were known to them alone. The source of these errors is that in one case functions were considered first and in another case, it was form.

It is then that a counsel that I gave, wrongly or rightly, intervened: instead of enlarging or strengthening the base so long admitted, my suggestion reversed it entirely, for it was a question of nothing less than a path diametrically opposed to it. In fact, if we are to act from a philosophical point of view in the investigation of analogies, I proposed rejecting considerations drawn from forms and from functions. Forms, I said, are ephemeral from one species to another; this view applies with further scope to functions, which grow in importance with their volume, everything else remaining otherwise in the same state.

Nobody thought of the inconveniences of practicing philosophical anatomy through the consideration of forms: it is to fall into antilogy. And in fact, to infer the facts of relations from the observation of differences was to accept judgments that rested on perpetual contradictions in ideas. However, I am far from blaming what was done at first; only this defective method was then known. That irregular progress was better than a baleful repose, for there is no worse condition for humanity than to be ignorant without suspecting it.

As to functions, we do not have to seek far for an example that justifies the exclusion we have made of them. In fact, what is there more perfectly analogous than man at his birth and the adult man? All the organs of progressive movement are in the one as they are in the other; it is the same with organs of prehension, of all of them definitively. But put them to work, and you find that what is easy here, is there impossible. The delicate hand of the infant will not lift this heavy hammer, which is the tool used at every moment by the hand of the blacksmith. However, in the identity of its parts, it is the same organ; its structure is the same, but its power is different because function follows the variable degree of every possible dimension.

Let us apply this to the comparison of the last part of the forelimb in the mammals. It will then be the case that without meriting the reproach that I have had recourse to an *emphatic* expression, I will be able to state my views as usefully generalized and concentrated under the name of *unity of composition*, or, to provide an explanation in a longer phrase, under the denomination of *unity of system in composition* of the parts being examined, for all my thought as expressed in those terms finds in this case a happy and complete instantiation.

In fact, the structure of the last quarter of the forelimb is the same: same use of phalanges, same adjustment and disposition to make them into fingers, same muscular apparatus to extend and to bend them. Why then do I not say uniform repetition of materials; do I not see there an incontestable identity of essence? All that is one and the same thing, whether in this species or in that; notice, however, that the function differs. For that last section of the forelimb is employed differently in a number of mammals, becoming the dog's paw, the cat's claw, the ape's hand, a wing with the bats, a flipper in the seal, and finally a part of the leg in the ruminants.

Now let us show that the theory of analogues is not at all a disguised repetition of the Aristotelian doctrine, that it is not a simple amplification of that doctrine, that it recognizes its own principles, that it has its precise goal, that it becomes an instrument of discovery, and that, in confining itself severely to the object under consideration for whose sake it exists, that is to say, an anatomical fact, it introduces into the study of animal systems the only scientific element able to make us apprehend all the physical conformities of the same rank.

1. It is in no way a disguised repetition of ancient ideas on the analogies of organization, for the theory of analogues forbids at the point of departure considerations of form and function.
2. It not only enlarges the ancient foundations of zoology; it reverses them by its recommendation of holding to a single element of consideration for its first subject of study.
3. It recognizes other principles, since, for it, it is not the organs in their totality that are analogues, which does however happen in animals that are nearly similar, but it is the materials of which the organs are composed.

This point is fundamental in my new doctrine. I will try to make it understood. An organ consists of the union of several parts, which, associated for a single purpose, simultaneously agree in the production of the acts and sensations of animals. This granted, if some of the parts composing it change by lengthening or by diminution, or even come to be absent altogether, that organ, well enough preserved as a whole, is nevertheless affected by a perceptible change. It is the same, if, without being entirely withdrawn, some parts of it are detached and joined to other neighboring organs.

But let us avoid all abstraction, and explain ourselves through examples. The human hyoid is composed of five ossicles, that of the cat, of nine. To the one as to the other of these organs the same name has been given, and, it will be said, with good reason, since the one and the other serve the same purpose. Are they analogous? The Aristotelian doctrine will answer in the affirmative on the ground that the two hyoids are very similar. The theory of analogues, on the other hand, refuses to see a complete analogy in this case, since there are more parts in one of the hyoids and fewer in the other. The latter approach must first satisfy the spirit of its own investigation, for it cannot judge with certainty until after it has retrieved the four ossicles absent in the human hyoid, or at least recognized the reasons for their total disappearance. Thus for the followers of the Aristotelian philosophy, it is enough for the function to be perceived; for them the whole organ, be it with five or with nine ossicles, constitutes an analogous organ. In a wholly contrary manner, the theory of analogues tries to discover, among the nine pieces of the organ as a whole, the analogues of the five reduced to that number; for it brings the analogy to bear only on the available materials.

4. Its precise goal is different, for it demands a mathematical rigor in the determination of every kind of material taken separately.
5. It becomes an instrument of discovery.

To show this, let us take the example we have just been referring to. In fact, the theory of analogues has to devote itself to the spirit of research: it has to inquire about the four ossicles, which, absent in the human hyoid, prevent that organ from being at its full development. If it has no reason to believe they have entirely disappeared, it will look for them close by, but outside the reduced organ; and if it wants to retrieve those anatomical elements without any hesitation, and find them again without difficult investigations, it will be able to fall back on another principle, its associate, its faithful guide: that is the principle of connections, a kind of Ariadne's thread, which keeps to the true path, and which necessarily leads to a happy outcome.

The hyoid of mammals, when it achieves its maximum composition, consists of nine pieces, arranged in two crossed chains, one longitudinal, fixed at the number of three pieces, between the tongue and the larynx, and the other transverse, composed of six, three at the right and three at the left.

With these hyoids in the grand totality of their parts, with this apparatus of nine ossicles, compare what has been preserved in the human hyoid. You find that the same materials are present identically in the two cases: to wit:

1. Among the pieces of the longitudinal chain and between the tongue and the larynx, there are on the side of the tongue, an odd bone, or the median arch: that is the main body of the hyoid; and in front, on the side of the larynx, paired bones, or the *large callosities* in man.
2. To form the transverse chain, there are now only the *small callosities*, the body in the center.

And in all, those are the five pieces of the hyoid in man: the median bone, the pair or large callosities, and the pair of small callosities.

Where does the theory come in, in this investigation? In establishing that in man the transverse chain is incomplete; that it is reduced, on the right and the left of the median body, to one ossicle, itself atrophied.

However, given the hyoid of the cat and of other mammals, would it be the case that in man, to form that number, there would exist outside and following the small callosities, consequently toward the skull, two other ossicles completing the transverse chain of which we have spoken above? Yes, that is what observation shows. Thus there are at the right two pieces, at the left two other similar pieces, which are lacking in the human hyoid. There is a cause of that effect: the upright posture of the species has produced that result. That is without doubt a grave anomaly, if we judge from the totality of mammals the rule for admitting animals to that class. The erect position of the spinal column, principally of the cervical vertebrae, which are its first part, and the very great size of the base of the skull are the hindrance that has kept the chain from being complete, and from being able to extend behind the ear, as occurs in other mammals.

I have named as follows the transverse chain when it has reached its complete number of seven pieces:

Stylhyal, ceratohyal, apohyal, BASIHYAL, apohyal, ceratohyal, stylhyal.

In man this chain is composed of the three following pieces:

Apohyal, BASIHYAL, apohyal.

We see in this table what extra pieces exist in the cat and in other mammals, which stand on their four feet, and are lacking in man, who stands and walks on this two posterior limbs.

But is the human hyoid nevertheless absolutely deprived of the lateral parts of the skull? That is not the case. A ligament coming from each small callosity or from the apohyal is prolonged laterally and reaches the extremity of the styloid apophysis.

But this is a new circumstance for the theory of analogues to investigate: the mammals, in whom this transverse chain is entirely osseous, lack that osseous thread characteristic of man alone. In infancy this thread is not yet attached to the skull; but it is nevertheless a bone of the chain, as in the mammals. But further, observation directed by the theory made me discover two primitive bones. To retrieve in man the stylhyal and the ceratohyal was the work of the theory of analogues, for all prediction is the fact and the goal of theories. Such successes confirm the utility of their creation.

It is not in this memoir, where I am dealing only with a general question, that I need to insist further on this special fact. Further developments, which I have already provided elsewhere, in the first volume of my *Anatomical Philosophy,* would here be superfluous. It is sufficient for me to add that human anatomy had already noticed and described the hyoid materials lacking in man; but it had not observed them as an annex of the skull. In its observation without doctrine, it saw there only a needle-shaped projection, fastened to the auditory tube, which would have been enough to give to that dismemberment of the hyoid apparatus the name of *styloid apophysis.*

6. Finally, the theory of analogues, to become everywhere equally comparative, confines itself in this case to the observation of a single order of facts.

It is necessarily exclusive in this respect. It cannot be at the same time anatomical and physiological. Before asking what this body is going to do, it is required that it be itself first established, that it be so independently of its form and its uses. All the advantages of the new theory are gained for it from the fact that at the beginning of its labors, it restricts itself to being anatomical, that it brings its examination to bear first and exclusively on the object under consideration, on that object as existing, and that it postpones for another study the investigation of its properties. Thus, with this unique element being solely under consideration, it is determined with rigor; it is not being followed in all its metamorphoses, and after it has been opposed to itself in all the cases, we come to know it analogically, that is, to understand it in its philosophical unity, without a mixture of any accessory consideration.

Let us once more have recourse to an example to establish this fact. Is it a question of a fingernail? The notions that the theory of analogies will concern itself with collecting with respect to that organic element are all those that concern its essence and its common properties, but this above all independently of its form and its uses, the consideration of which has importance only relative to each species separately. The different sizes that it can acquire, if there is no question of a particular fact, and if we are to confine ourselves to the most general point of view, could not form for us a consideration of any interest whatsoever. For, whether it is a thin, small epidermal cap, as in the *unguiculates*, which is then called a NAIL, or becomes a thick horny mass, as in the horses, the ruminates, the *ungulates*, a mass to which in this case usage assigns the name of HOOF, the theory of analogues, seeing the caps of the last phalange of the digits from its philosophical point of view, considers them as one and the same thing; it makes no distinction between them.

In the second phase, we come to the study of forms and functions, attention is transferred to the metamorphoses of these identical elements, to admitting their different sizes and to recognizing their different uses.

This is not only a point of theory perceptible to the eye of the mind. Nature, in its errors, which we call the facts of monstrosity, gives us a complete demonstration for the eyes of the body.

The rule, that is to say, the general fact, for all animals with four extremities is shown in the digital subdivision of the last part of the limb. If there are five distinct digits, the five digits attain only an inconsiderable size and are respectively more or less the same: the nails are also small, and hence of a size to be only nails, according to the strictest acceptance of the word. But if it happens that the lateral digits are sacrificed, because most of the nutriment goes to the intermediate digits, as in the ruminants, in whose case two digits develop with hypertrophy, while the two others remain in a state of atrophy, or else, as in the horse, which is in the first case for one digit, and in the second for two, the nails experience the same overdevelopment, and become thickened nails, or *hoofs*.

The foot of a ruminant, and still more that of the horse, are exceptional cases, are what in deference to the inclination and the habits of behavior of our minds, we speak of as, we call, cases of anomaly. It is in these circumstances that I have seen the rule take over again in some monstrous horses. The upright and learned M. Brédin, director of the veterinary school of Lyon, showed me a horse born with three toes in front and four toes behind. Returning to Paris in 1826, I published that fact, recalling that there were others in the annals of science, namely: a didactyl horse which, according to Suetonius, lived in Caesar's stables, another similar horse that belonged to Leon X, etc.

But in all these horses, which monstrosity thus returned to the usual rule, to the plurality of digits, the nails remained thin and small, true nails in the strictest sense of that term.

In this first memoir I confine myself to general considerations that I have been presenting, and, I assure my readers, I have barely broached a subject of inexhaustible fecundity. I have said nothing of my work on the skull; of that destined to relate the fishes to the organization of animals that breath in the air; and nevertheless, those are studies that have recalled a number of rules, some of which have not yet been mentioned here.

That is where all the difficulties of the subject were: to disentangle them was to exhibit new relationships; it was, through acquiring such results in order to compose of them the domain of science—it was to generalize; that is to say, to embrace all the particular truths in common and elevated considerations, of which the theory of analogues is definitely only one expression.

In the following memoir I will enter further into the depth of the question. There I will examine that particular point on which our debates bear, namely, if necessary, with Baron Cuvier to restrict further and further, or, on the contrary, with me to augment further and further the field of philosophical considerations.

It is enough for me, today, to have shown that I have corrected, renewed, and made more precise ideas on the analogy of organization, and that I have substituted for the nonscience of reigning opinions a clarified and certain course, which has become a method that can guide us usefully in the strict determination of organs.

In conclusion, I regret, and I express all my sorrow that there has been a conflict of opinion between the oldest of my friends and me, but I could not keep myself from raising my voice. For it is not at all for my own advantage that I have done it, but to offer the development of a doctrine that I believe has at present achieved a high degree of utility. Since I have acquired, by the employment of my life and the pursuit of a single aim, the character of Saint Augustine's *homo unius libri* (*man of one book*)—an expression which that learned bishop applied to those who advance one principal idea—, I have taken the opportunity that has been offered me to explain how I understand that one thing of which I ALWAYS DREAM.

ON THE THEORY OF ANALOGUES

Applied to the knowledge of the organization of fishes

(SESSION OF MARCH 22, 1830)

The system of argumentation that was opposed to me at the session of last February 22 consisted of two distinct parts: of the two following objections.

FIRST OBJECTION: *If, in insisting on the analogies of beings, you confine yourself within strict limits, you are only saying something true, agreed on for 2,200 years, and first proposed by Aristotle.* I replied, in the memoir that I read the first of this month, that my doctrine called *theory of analogues,* resting solely on the doctrines of anatomy and in all respects on different principles, was in no way a repetition of the Aristotelian doctrine.

SECOND OBJECTION: *To arrive at a principle of unity, you leave the field of genuinely comparable facts, you give it a scope which ought, on the contrary, to be restricted, in order to confine it within narrower limits.* It is this point that I am going to examine today insofar as it concerns the organization of fishes.

Later, I will examine in addition the value of this objection; first, in what concerns the anomalies of organic development in each animal, anomalies that constitute the facts of monstrosity; another time, by giving a summary of my research on the composition of the bony head; and finally, in a third memoir, by recalling what relations science has found among the higher animals and the crustacea, the insects, and in general all the articulated animals.*

*I flatter myself that I am continually concerned with these considerations; as I understand them, they will consist in a revision of my former views, to which I shall have much to add. Too extensive in their subject matter, no one of them was able to find room in this first publication.

In anticipation, kindly allow me to note here the idea of a very advanced report. I am doubtless risking a great deal, by depriving it of the support of facts indispensable to its development.

The insects and the mollusks, if we take as typical the beings at the center of each series, are very different, and present above all important traits to be established, still less for

Must one in fact force oneself to extend further and further, or ought one rather, on the contrary, to hold within restricted limits the applications of the principle of the philosophical resemblance of beings? I do not intend to treat this question today except in what concerns the organization of fishes.

However, before embarking on this subject, I sense, and in no way wish to avoid, an objection that could be made to me, and which I here state as follows: "It is of mollusks and not of fishes that it was a question at the beginning of these debates; to refuse to enter, even at this moment, into the territory of the conflict is to open oneself to the inevitable prejudice of a position already taken, to the weight of a decision firmly pronounced and that is consigned to science in the following manner: *The cephalopods are not the passage to anything, having in no way resulted from the development of other animals, and their own development having, in no way either, produced anything superior to them.*"*

their extreme precision than for a very curious character of reciprocally inverse relations. For, further, if you judge the two embranchements on the animals at the confines of each series, you see them enter into a common conformation, and blend to such a point that the limit of demarcation between the two great families is difficult to place.

The composition of the animal is ascertained usefully only at the middle of a proportional, regular, and harmonic distribution of the two principal systems, one for the circulation of fluids, and the other for nervous excitation. It is clear that in the successive and progressive developments of organization, the blood system and the nervous system stand to one another in a necessary relation. However, it is for observation to determine to what degree. Now what every one has been able to notice as a particular fact, what every one will at once find himself cognizant of as a general pronouncement, is the respective position of these two systems in the insects and in the mollusks; it is their inverse balancement as to quantity, from which each group receives its special character. The blood system is excessive and, on the contrary, the nervous system is seized with atrophy in the mollusks; it is the inverse with the insects. This explains the large hiatus that has been noticed between the families, especially with regard to the beings at the center of each series, and also the very numerous relations that they exhibit at their extremes; for, if there are mollusks with the nervous system proportionately well developed, and insects similarly in excess with regard to the blood system, these are so many conditions that converge toward the same point, to lead to a common conformation. But this *coming* and *going* of an organization, here richer and there much less so, furnishes facts for hiatuses that are more and less large, without compromising in any way at all the principle of the unity of organic composition.

*I have been reproached for seeking detours to avoid replying categorically on the cephalopods, the true terrain of the controversy, according to a number of people.

First: I have explained about the intention I had to allow the young authors of the memoir on the mollusks the trouble and the merit of a reply.

The theory of analogues derives from its rules a character of inspiration and of the future. The dogmatic tone, applied to the judgment of differential cases, is especially repugnant to its ways. If, when it had not been employed until now for the determination of the organs of cephalopods, it has been silent with respect to them, would it be fair to use this for a definitive condemnation? No, certainly not. What does it matter if no facts have been acquired except for the following results, which I freely acknowledge? The cephalopods, which occupy an elevated rank among the lower animals, have not yet been studied except from the point of view of large intervals, of their distance from the groups to which they come closest. If, then, there are no other antecedents with respect to them, science alone is to blame; nothing had yet established that, on the questions that have been raised, the future of the theory of analogues, insofar as it touches the mollusks, has been the very least bit compromised.

How much hope is there, on the contrary, that in the future the true affinities of the mollusks will be finally expounded and explained! For this, nothing is needed but to pursue, by another path and to a suitable degree, investigations according to the spirit of our new method for the determination of organs: especially that nothing be demanded of the facts but their possible function relative only to the degree of organization at which they are observed. For it is a question of animals some degrees lower in the zoological scale, and consequently that amounts to considering beings that belong to one of the ages of variable developments of organization. And in fact it is correct to consider the mollusks as forever realizing one of the lower degrees of the progressive order of organic developments, to see them as stopped at one point, and, to that effect,

---

Second: I am here establishing the fact that I cannot excuse myself from studying in the first place the organization of the fishes.

This is not at all to refuse combat on the terrain of the mollusks. If the field is still free when I produce my second publication, I will have made investigations, and I will then offer them with perfect security.

However, if I have to act at this moment, a remark would suffice to show how false was the scaffolding of the arguments and the outlines they thought they could rely on. Everything rests on the following objection: "We would admit rigorously the hypothesis of MM. Laurencet and Meyranx and the comparison to which it gives rise, if it were only that they place the *brain* ahead of the neck."

After the fine research of M. Serres on the nervous system of animals, we know that there is no spinal cord or *brain* in the mollusks, any more than in the insects. For an instant I thought and said the contrary; the argumentation took note of this. Granted, I was then, like all naturalists, under the sway of opinions and errors of the infancy of the science. I was wrong; I recognize it without pain.

as not yet having furnished a certain kind of organ, or if the latter does begin to come in sight, as not yet having produced it in full strength.

That is why I should not, and cannot, compare them at once with one another at the extreme ends of the scale. I would first have to give all possible attention to the intermediate links: otherwise this would be to take a backward step in the logical order of ideas, and in fact to begin where, on the contrary, it would be appropriate to end.

That is what I would be doing in the same manner, if I had to demonstrate that the bud that appears first belongs, but with an inferior degree of organization, to the same system of composition as the branch that will arise from it. And let us apply this, for example, to the bud from which there will grow the vine-stock, richly charged and ornamented with hanging grapes. It would be no more reasonable, or logical, to try to provide an explanation in regard to this, while omitting the examination of all the intermediate ages of the branch or the consideration of the successive degrees of development.

It is the same for every family considered in the middle degrees of the scale: each one corresponds to one of the ages that the bud has to pass through, in order to come to produce its branch and its fruits at full strength.

That granted, we have no way of evading a given situation. The fishes come after the reptiles and above the mollusks; therefore, the fishes are necessarily that intermediate link that the logical order of ideas already calls on us to examine.

But first, what do we find established in this respect in Aristotle, in the works of the founder of comparative anatomy, source invoked for all enlightenments? What confusion exists there! We shall see. The mollusks are not fishes, his history of animals tells us, in book 4, chapter 1, because they have no blood; then, further on, in book 9, chapter 2, it is said that they do form part of that class. At least Aristotle ranks the squid among them; seeing in that reference only the effect of a misunderstanding, I conclude from it at least that Aristotle thought mollusks were placed close to the fishes.

Another question merits a little more attention, that of knowing if the fishes were, with regard to their material constituents, reduced to the animals by which they were preceded, and with which they have always and forever been classified. If there is here still a fact left in question, one can understand that it will have to be treated first, for we could not leave this point of the discussion behind without depriving it, for the sake of the general question, of the most necessary facts and of the action of their power. And in fact, if you passed from the fishes legitimately included in the embranchement of the vertebrates, from the fishes completely reduced with respect to the higher animals through the identity of all the details of their organs to beings of the second series that come next, this is to procure the support of natural transition, this is to control a future in order to know better those animals of a lower degree, which, it is said, a manifest hiatus separates absolutely, and which must consequently be attributed to another plan.

Let us here make one remark; it is that, if the conflict that is taking place today had had the fishes for its subject matter fifteen years ago, we would have been caught unawares much more seriously than we are on the subject of the mollusks. For at that time no one had ventured ex professo on the path of the philosophical determination or organs. But at present, in place of a kind of groping and of the resources of an instinct more or less well directed, we possess a body of principles in the theory of analogues. Thus, fifteen years ago, one would quite naturally have said, one would have easily established, founding oneself on the Aristotelian doctrine, that there was no appreciable and precise relation between animals with atmospheric respiration and those with aquatic respiration, with respect to their respiratory organs. Indeed, a clever argument, mastering the scientific facts as they then existed, without being held up by the decisions of specialists in method, by the givens of classification so far approved, would not have failed to pronounce in favor of the existence of a separate ichthyological type. For someone who was studying the organs of respiration fifteen years ago, the differences were everywhere, while the analogy of constituent materials appeared nowhere.

But at last, after the period in which the facts were studied from the point of view of their differences, there came that of the investigations of their relations; I have employed fifteen to twenty years in these investigations with respect to the fishes; and it was late enough that I came to think, to admit with all confidence, that there are no materials created especially for an ichthyological type; and that, consequently, there exist in the fishes, as in the higher animals up to man, for the composition of the respiratory organs, only a certain number of identical parts, which are, essentially speaking, absolutely the same, but which, susceptible of variation in their respective volume, draw the ground for their modifications as forms and as functions from the influence of the environment in which these same parts are called on to develop.

I shall make my thought palpable by citing an example that everyone can appreciate. The rose that has kept its stamens interests the botanist because of the maintenance of the reproductive power, and the rose that has lost them, through a transformation into petals, simply gives more pleasure to the gardener whose flower-beds it embellishes. But for the philosopher who evades the conclusions of those special positions, these two kinds of rose are but one and the same plant, variable under the influence of surrounding environments; for this member of the Rosaceae is composed of parts that are the same in substance, identical as constituent elements. The form and the functions of these parts are of no importance from this point of view; only, as the influence and the reactions of its external world determine, this element is either a stamen or a petal; but before any acquired quality, each element is first itself, then capable of all possible volumes, that is to say, capable of maintaining itself in a *medium*, of restricting itself to a *minimum*, finally of being carried to the *maximum* of its development, sometimes to the point of suffering the extremes of the most bizarre metamorphosis.

What has been so cleverly combined since the origin of things, that it is acceptable to oppose to us the *consensus omnium*, which seems to give to the determination of organs their present names? What would the Aristotelian doctrine, several centuries before the Christian era, have in fact established, so that we must avail ourselves of it today and so that it would be right to declare that it must be adhered to? There is nothing real in favor of the past except for one sole reason, which is not a good one, and that is that the ancient usages have been for so long not subject to revision, and that people have been for a long time satisfied with opinions that nevertheless have not always remained stationary. We are the first to announce that, during the centuries, and chiefly through the labors of the author of lectures on comparative anatomy, a very extensive knowledge, the resources of an exquisite wisdom, and the good fortune of laborious efforts have brought to light a great number of precious relations, all unperceived in the infancy of the science. The labors of Perrault, of Daubenton, of Vicq d'Azir, but particularly those of 1795 and the years following, have begun to make of comparative anatomy a positive science.

However, what were the inspirations and the procedures of Aristotle? How did he understand both the relations and the differentiated traits of beings? I distinguish, he wrote, *two kinds of animals, those that have blood, and the others, who do not have it.* This division and the idea on which it rests have always been reproduced; in the time of Linnaeus, the talk was of *animals with red blood and animals with white blood;* Lamarck recommended, and brought about the adoption of, that other formula, *vertebrates* and *invertebrates.*

For Aristotle, there were thus animals of two kinds; but notice: he did not say, of two types. On the contrary, he made them issue from a primordial type. According to this philosopher, there are first animals: considering them thus abstractly, he takes that general view for a first fact, and it is only secondarily that he perceives in them distinct qualities. Thus animal organization is founded, in the Aristotelian conception, on something essential and primitive which, unfortunately, he did not specify by adding: on a single system of composition for the organs. We are completing his thought.

On this first part of Aristotle's views, we do not differ at all: consequently, the priority of his views is granted; but as to the second point of his ancient doctrine, we differ from him entirely. For want of understanding that this composition of organs, basically one, essentially the same, as residing solely in the consideration of the anatomical element, was alterable to any degree whatsoever through the action of the external world, the Greek philosopher believed that the analogies of organization, intuited and perceived by his genius, rested entirely on the consideration of forms and functions. That is the error introduced into his doctrine, an error that has been perpetuated through so many centuries. It is this error from which we are protecting ourselves today through the theory of analogues, which, being established through a true principle, has since caused so many disagreements. This principle, vitiated in its application, and the error that obscured its useful reflection, worked at the same time to inspire equally both the naturalists who held to a reality of absolute differences, and those who claimed to assemble and to reduce the facts of variation to the unity of relationships. Such are the confused ideas that have more or less profoundly penetrated the works of the preceding school, and of which one can find a remarkable example in the following passage: "*There are no resemblances between the organs of fishes and those of other classes, except insofar as there are [resemblances] in the functions*" (Cuvier, *History of Fishes,* vol. 1, p. 550).

Absolute resemblances, without the least doubt; who could doubt it? However, as it is placed in that phrase, the word *resemblance* is equivocal, since it can be understood in one case as philosophical resemblance, then in another strictly as perfect similarity.

At this moment it would perhaps be useful to the development of my thesis if, through an historical summary of what had been practiced, I brought out all the inconsequentialities of the usual procedures in the impositions of names that were attributed to supposedly identical organs. There was difficulty in operating when it was a case of passing from a well-known family to another placed at large enough intervals from it. The considerations drawn from the form and the function formed the point of departure; the cephalopods and crustaceans clamber or creep on the surface of the ground; the appendices they use for this are taken as feet. In the decapod crustaceans, these appendices, the very same ones, essentially speaking, are of three kinds as far as their use is concerned; the front ones are used to seize food; those in the middle for walking; and finally the back ones call to mind only the idea of their uselessness, whether for locomotion or in any other way. Now such are their names: *maxillipeds (*that *is, jaw feet), walking legs,* and *false feet (swimmerets).* Thus function is always placed at the forefront of the considerations invoked: that there is this leitmotif for deciding is enough to arrive at a common name; to doubt in order to make a better judgment on this consideration, to justify this point of view, would not be to attend to what is most urgent; it is enough that the function presents itself under a new aspect. In this case, no one has any scruples; new names intervene. Thus, for several of the materials of the organization of fishes new names have been imagined, names that were not customary for other vertebrates, like *opercular* or *gill cover, preopercular, interopercular, subopercular, branchiostegal rays, branchial arches, gills,* etc. However, the function invoked in this case—the function signifies only use or service. But then I ask: use, service of what? What part of the body would have, would find itself having, this function? What are these materials intrinsically, in themselves? That is what you leave, without even having thought about it, among the unknowns of your problem: you have so far given to the object of your considerations only a temporary name.

But no ichthyologist has thought of making this precise statement, which should have the advantage of presenting the actual state of science. From this it has followed that every omission in this respect is equivalent to a declaration, implicitly announced, that there are in fishes some materials that deviate from the common plan, created especially for them, and, finally, that it is the novelty of these parts that has caused the recourse to new appellations.

But as to such a specialization, I contest it formally; moreover, I hold it to be impossible. And in fact, when the fishes correspond to the higher classes in almost all their organs, this would be to admit that on one single point, their respiratory apparatus, this correspondence would be lacking. To make such a supposition: is that not to believe possible the alliance of heterogeneous things? Is it not really to withdraw its principle of existence from an organic composite, which is and can be only through the reciprocal relations and the harmony of its constituent parts?

But let us stop worrying about what was done in the infancy of science, of what there was that was misguided in the terms that were used to express ideas not yet sufficiently elaborated; and let us look at our subject from a more elevated point of view.

There is no animal organization except through the necessary intervention and under the power of the phenomenon of respiration. But the exercise of this phenomenon is possible only in two different mediums, air and water. With the differences of their respective densities, these two fluids would have been equally able to receive different conditions of existence and, for example, to find themselves working with a complete and reciprocal independence relative to the animals in question.

I did not at first think of that hypothesis; my first investigations were made under the inspiration of Aristotelian ideas. But arrived at the middle of my career, I thought it necessary to turn back to this point, and to examine fundamentally the question of knowing whether the two mediums, of which I could not fail to recognize the powerful intervention, all the force of reaction, could either have the power to demand that animal organization should be provided in advance with conditions of a distinct kind, or would rather find itself sufficiently adapted to conditions of existence of a single type, in this case preexisting before any function, but that each milieu would have the resources to effect modification, that is to say, to adjust to the character of its specific density.

Lacépède must have believed in the first of these hypotheses and assumed the action of a double primitive given, and thus in fact have considered animal organization as subjected to the development of two distinct plans, when, in the preliminary discourse of his history of fishes, he proceeded to propose a new theory of respiration for animals provided with gills. According to the principles of this theory, it is water as such, and in no way the air disseminated among the molecules of water, that fishes breathe directly. The decomposition of the water would be produced by their vital action; a distinctive mechanism, or another sort of respiratory apparatus, would have this power and would give this result. On the given hypothesis, we follow the two elements of the liquid after their separation; each one is incorporated in its own way in the substance of the organs. However, it was not found that the effects corresponded, as to the degrees of their differences, to the diversity of the cause. Beings developing under the influence of such a regime ought to justify it by forms yet more singular than are those of fishes and ought to yield products that were altogether bizarre, outstanding conformations exceeding all predictions, all the most exaggerated suppositions.

When the facts were examined, the second hypothesis seemed the correct expression: no one today doubts this. Thus there would be, there really is, only one system of organic composition, only one primitive design to arrange the regulation of its parts, only one single plan, in short, unique with respect to what forms the essence and the interconnection of the elements comprised in all organic formation. But this system is alterable in its parts, on the part of the ambient milieus, from which it derives the assimilable elements and the ground for the variation on every point—a difference introduced through the diversity of their respective volumes.

What facts would have yielded these responses with so much precision? What investigations authorize me to be entirely confident here? To explain this, it will suffice to narrate what happened to me. From 1804 to 1812 I acted under the inspiration of science as it then existed. Describing, for the great work on Egypt, a fish of the genus *Tetrodon*, I had needed to determine a piece of an extraordinary size, which plays a very remarkable role in the behavior of that species. It is a long bone, which takes the place of the absent ribs. Reaching to it and attached to it, on the one hand, are the muscles of the shoulder and, on the other, the intercostal muscles: the former push it forward, and the latter backward: a variable position, to which the curious phenomena of the distension of the tetrodons is related, and by means of which they pass from an elongated form to one that is entirely spheroidal. This bone, on whose existence depend so many curious facts of individual activity, needed to be called by its name; but this name was lacking. Instead of creating it for this particular and arbitrary circumstance, I preferred to inquire of science and to derive it from deductions of analogy, and it is from that period that my first investigations of the philosophical resemblance of organs are dated. I fastened on the idea that it was a part of the shoulder, and I made up my mind to call it the *coracoid bone*.

With that piece thus dealt with, I passed to the adjacent pieces, applying myself to running gradually through all the anatomical regions. In terms of forms, this was for me a new spectacle, for nothing, or almost nothing, of the appearance that other vertebrates exhibit was preserved in the fishes. To the degree to which the difficulties multiplied, I had the hope of triumphing over them through dogged work, when I found myself brought definitively to a halt.

That happened when I faced this question: *What is the gill cover?* What part of the organization of the higher classes could furnish its analogy? From 1809 to 1812, I made futile efforts to find this out. After many hypotheses, which turned out to be false speculations, I gave up; I halted before this obstacle, which I decidedly considered insurmountable.

Thus my investigations, at first so ardently pursued, were no longer invigorated by the principle that had inspired them. There was no more hope to apply them to the totality of organs; and what made this crisis yet more painful was that the obstacle that stopped me made me doubt the reality of the relationships I had previously found. After so much useless labor I abandoned my ideas only with a very lively feeling of regret. However, in 1817 an intellectual awakening made me aware that the five years of my involuntary pause had not passed by without bearing fruit. I believed at last in the solution to the question, *What is the gill cover of the fishes?*, when I came to understand that the three bony plates of the gill cover are analogues to the chain of ossicles, called specially in man and in the mammals the auditory ossicles of the inner ear.

From this moment I again took courage and recommenced my work, never again to abandon it. My ideas, formerly fixed, grew in extent. The very obstacles that had stopped me, when examined for their implications, were appreciated. In directing my thought to the mistakes I had made, my recollections became for me such a useful source of instruction that, engaged in deep meditations about them, I was led insensibly to the interconnection of the facts: once I had brought them all together, I saw them finally leading to high and important generalizations, to the establishment of certain rules and to the revelation of principles that are the foundation of my theory of analogues.

One can now understand that, relying on such support, on a theory thus deduced from a great number of facts and of general propositions that furnish their justification, I am no longer astonished at the transformations that the parts employed in the act of respiration undergo. Necessarily at bottom the same, since they have to exist in harmony with the other organic systems, whose mutual relations are in no way contested—necessarily, I say, the same at bottom, they arrive at precisely the state of transformation in which one would have to expect to find them. For they should be, and they are in fact, modified and adapted to the diverse nature of the two mediums, air and water, in which they are called on to become active. This would itself be an inexplicable fact, a cause lacking its effect, if the parts of the respiratory organ did not respond by a variation of its form proportional to the difference in the density of the two milieus. That is how the great metamorphoses of the respiratory parts became for me only a simple fact, only the consequence of perceived premises.

That given, I asked myself what would happen to the materials employed in the play of respiratory phenomena if they entered successively into functioning in the two mediums, and I found that the fact itself replied peremptorily. It is necessary, in fact, (1), as to the atmospheric milieu, that the surfaces of the apparatus increase, that it gain in length, and that it establish itself in the center of the animal; for the elastic air can insinuate itself into the deepest recesses if an issue is contrived for it for that purpose; and (2), as to the aquatic milieu, that all the parts of the apparatus approach one another, that they be concentrated and led to the outside of the animal, so that they can be constantly submerged in the ambient milieu—a liquid without buoyancy, in which every molecule of blood has no longer any resource except an immediate contact for overcoming several forms of resistance, the cohesion of the air with the water and that of the two elements of the air themselves. Now that is what a posteriori investigations pursued during twenty years of my life have made me recognize as being what exists, as providing in reality the relation of animals to their ambient milieus.

Yes, without the least doubt, the whole respiratory apparatus is modified into only two systems:* the forms that these two systems assume and the function they fulfill vary just as do the resistances of the ambient milieus; but the apparatus, as to its essence and the arrangement of its elements, remains basically the same. And does it not in fact appear obvious that this is a unique apparatus, that it is, in the last analysis, to an identical organ that it belongs to produce what is in the two cases only the same phenomenon, a phenomenon that consists in the combustion of a part of the blood through the absorption of the oxygen of the air?*

*Two weeks after the reading of this memoir, the Academy received from M. Flourens a communication in which the mechanism of the respiration of fishes is more ingeniously presented and explained. Functions reduced to similarity of action seem to be the chief aim of that work. This coincidence has struck a number of people.

*My son (ISIDORE G.S.-H.), dealing, in the large work on Egypt, with the silurid harmout, a species of fish of the Nile, brought out the following remark, seeing it as following from my former investigations:

> Animals possess elementarily two respiratory apparatuses, the one *branchial*, rudimentary in the species that breathe in the air, highly developed in those that breathe in the water; the other *pulmonary*, rudimentary in the species that breathe in the water, and highly developed in those that breathe in the air. To the first of these divisions belong essentially the mammals, the birds, etc; to the second, the fishes and several families of invertebrates. But the two systems of organization that these two divisions present are not the only ones that can be encountered in the animal series: for just as there exist beings that have the power of breathing both in an atmospheric and in a liquid medium, there even exist beings in whom are found at one and the same time a middling degree of development of both the pulmonary and the branchial apparatus: such are a number of reptiles, like the siren, the proteus, and the tadpoles of other Batrachia; and such appear also to be a number of other crustaceans, and particularly the genus *Birgus*. These ideas, which my father communicated to the Academy of Sciences in September 1825, led him to consider, in the silurid, the organ for-

In a vague way the methodical naturalists have seen this and have declared it implicitly in their classifications, when, without the least hesitation, they placed the fishes in the embranchement of vertebrates. But in agreeing to these views of the relations in question, would not those naturalists have surrendered only to the need to align, to adjust, and to isolate beings in their classifications? One is truly tempted to believe it, since barely has their work borne some fruit, but this is at once denied in the execution. They soon distinguish in the fishes between the parts that are reduced to their analogues in the higher animals, and others that are not so: the former have a common name, and the latter, on the contrary, a special name, as if they were a new product of the Creation.

Let us explain this. There can be no doubt: no one was filled with jollity at this manifest contradiction; they were driven to it by the need to progress rapidly in the work of ichthyology properly speaking. Zoology, in its need for activity, did not wait for the more reflective and slower work of zootomy. The latter was not able to deliver in time its philosophical considerations. Names were necessary, they had to be provided. Provisional names were thus imagined and accepted, to assist in describing the species. If that is the case, this provisional establishment by no means constitutes a legitimate official possession and ought not to be invoked as a result presenting the final achievement of science: this adoption of a special language attests only to uncritical habits.

In *The Natural History of Fishes,* recently published, the parts of the head of the fishes are reduced to their true analogues, on my view, only in a little more than a third of their number, thirteen out of thirty-two. The difference in point of departure explains how there can be so great a disagreement. In the opinion according to which only thirteen pieces are so reduced, the relations are admitted that bear on the object, its forms, and its functions; in the contrary system, which holds that the determination of thirty-two pieces is possible, consideration is given solely to the anatomical element. I return to the preference that I have been obliged to give to that unique point of view, in order to remark that to act otherwise is to recognize in the fishes two distinct natures: one bearing on the common organization of vertebrates, and the other attempting to escape entirely from that consideration. It is no longer possible to say that the determina-

merly designated under the name of *supernumerary gill* as an atmospheric organ of respiration, as a true lung. And it seems in fact, not only that the harmout can live several days outside the water, but that sometimes it even leaves the river on its own, and advances by crawling in the mud of the canals that issue in the Nile.

tions of organs, that all the efforts to reduce them to a single conformation, are improbable for the reason that they have not been discovered, and that it is useless to try to discover them. I may recall that the first volume of my *Anatomical Philosophy* was consecrated to showing that, part by part, there is no anatomical region that does not offer the character of the philosophical similarity of organization, that is not in fact decidedly reduced to their common relations.

This whole discussion specifies the point of our controversy. The field of philosophical considerations is necessarily restricted in the case where three elements that do not always coincide with one another are called on to compete; and quite the contrary, this principle becomes a subject of indefinitely extended observations, resting solely on the consideration of the anatomical element. In the first case, it is all at the same time the subject, its forms, and its functions, three conditions that cannot meet and do not meet in a united fashion except in animals of a single class; in the second case, the anatomical element remains comparable everywhere, even when it disappears—since then there remain, even for observation, traces indicative of its disappearance.

But there is something still better, and it is with this last reflection that I shall conclude: the function itself, when it is included in its general formulation, is never really absent: it is recovered in its entirety in the cases I have been referring to. In fact, where do the differential facts strike us? It is only in the regions and the parts, of which the whole is called the respiratory organ, the parts here accommodated to the atmospheric milieu, and there to the aquatic. Let us look at the function: what should be definitively the employment and the use of this assemblage of pieces? To produce the oxygenation of the venous blood. But it is to this that the two sorts of respiratory organ both apply themselves. And indeed, in one case the air rushes down to the base of a pouch irrigated by blood; that is what the whole pulmonary apparatus consists of. And in the other, that same pouch, which loses its condition of being a pouch with only one opening, because it is pierced a number of times at its base, nevertheless reacts to the air caught and retained between the molecules of water: this organ, thus transformed, goes toward the respirable element, behaves as if it has been pressed back or thrown back, pushed back outside like a folded glove finger; in this other form it is called a *branchial apparatus*. Thus, even with regard to what concerns the functions, if we judge from above and with the definite goal of considering organization, the analogy is preserved.

From the facts put forward in this memoir I draw the conclusion that the questions of the philosophical resemblances of beings ought not to be confined within the limits set by the argument of February 22, and that consequently I could, and had to, extend the ideas of identity, the facts of the analogy of organs, in a sense broader than had been done before me.

And definitively, it is to give that same thought a more general expression, to consider that the hour has arrived for a salutary reform in the study and the language of the facts of animal organization. Would it in fact be wise to claim that we must always let ourselves be governed by insufficiently justified habits, that we must yield to the needs of the moment only by inspirations that are groping or conceived in ignorance of the facts, and, in short, to prefer vagueness and the oscillations of a doctrineless past to the teachings of the present time, rich in facts raised to the level of philosophy? On the contrary, we should rely on the inductions of so many novel general propositions, the totality of which become a sort of method, as furnishing the support of an assured guide, and as being truly an instrument of discovery that can be usefully applied to the determination of organic systems.

In other words, must the idea of a new scientific era in what concerns animal organization be rejected, or, on the contrary, admitted? Must we remain irrevocably in the paths of anatomy, successively and so diversely worn away, or rather try to open new paths, under the instruction and in the direction of recent discoveries?

SECOND ARGUMENT

BY BARON CUVIER

(SESSION OF MARCH 22, 1830)

The young writer, editor of the scientific section of *Les Débats*, opens the article that he has inserted in the March 23 number of his journal with the following reflections:

Many people are still asking what is meant in natural history by *unity of composition*, *unity of plan*. It is true that these rather vague words have never been well defined; but no doubt it will not be long before this happens, thanks to an unforeseen circumstance that must lead perforce to a clear and positive explanation on the part of two men equally interested in defending their opinion. The one, like Aristotle, applying his genius to the observation of facts, has elevated the monument which that great man had established on foundations that up to the present have been unshakeable; the other, full of imagination, wishing to open new paths to zoology, has embraced nature in an abstract and philosophical theory. We shall follow them with pleasure in a discussion where truth must finally emerge; we will abstain from injecting our own reflections, since we can do no better than to place before the eyes of our readers the parts of this proceeding.*

We are persuaded that they will understand the question perfectly after reading the following memoir, which M. Cuvier presented to today's session.

*The arguments that tend toward the condemnation of my ideas are almost the only parts of the procedure that have been placed before the eyes of the readers of the *Débats*, and that was inevitable, given the present editor of the science section. Having neither the knowledge of the subject matter nor the discernment necessary for undertaking an extract, he limited himself to bringing to the publishers' workmen the memoirs that had been entrusted to him; they drew on them according to the demands of the space available.

"Our learned colleague, in his last memoir, began by admitting with great honesty that by *unity of composition* he did not mean *identity of composition,* but only *analogy,* and that his theory should rather be called *theory of analogues.* So here is a great step in this matter. These equivocal words, which served only to confuse the ideas of beginners, *unity of composition, unity of plan,* will disappear from natural history; and if I had rendered only this service to science, I would already believe I had not wasted my time.*

"But nevertheless our colleague assures us, at least so far as I have been able to understand him, that his theory of analogues is something special:

"1. In that he neglects forms and functions in order to attach himself only to the materials of organs.

"2. In that the analogy resides solely in the identity of the constituent elements, and that this analogy knows no limits.

"On the first point, I shall not much insist; at bottom, it does not matter much whether a doctrine is new if it is false: I shall say only that I do not know one single anatomist, not a single one who had determined organs uniquely by their functions, much less by their forms. Certainly no one has yet been so bold as to say that a woman's hand is not a hand; and yet two weeks ago I would not have believed that any one would dare to say that a woman's hand did not fulfill the same functions as a man's; but those are assertions that escape in the heat of the dispute, and on which an adversary in good faith would have the generosity not to insist.

"What is certain is that the anatomist against whom the attacks have been chiefly directed, so that finally he sees himself with much regret obliged to repel them, is one of those who have had the most frequent occasions to make it plain that the functions of a given organ change according to the circumstances in which it is placed. But I repeat, these issues of vanity are of little importance. What interests the friends of truth is to know if this theory, which its author calls theory *of analogues,* is *universal,* as he says, or if, as other naturalists think, there are analogies of every sort, but always limited, and what are their limits?

"But how to discuss a question when one is not willing to state its terms?

*I have not made, nor have I had to make, any concession; I limited myself to declaring inexact certain phrases and certain confusions of ideas that had been attributed to me.

"In this regard I have made clear and positive demands. You attach yourself to the elements! Very well; do you mean that there are always *the same elements,* do you mean that these elements are always in *the same mutual arrangement;* finally, what do you mean by *universal analogies?**

"If our colleague had made a clear and precise reply to my questions, that would be a good point of departure for our discussion; but in his long deduction he has never replied, for it is not to reply to say that all animals are *a product of one and the same system of composition;* since that is to say the same thing again in other terms, and in terms that are much vaguer, much more obscure.

"There would seem to be a more positive reply in these words: that *animals result from an assemblage of organized parts that are uniformly repeated.*

"But push a bit on such a reply; you will see that if you take it literally it collapses of itself. Who will dare to tell us that *the medusa* and *the giraffe,* that *the elephant* and *the starfish*† *result from an assemblage of organic parts that are uni-*

* *Analogies universelles.* I wrote nothing like that. The associated terms make nonsense.

† This objection concerning *the medusa and the giraffe, the elephant and the starfish* caused a good deal of surprise, and will, I believe, cause even more in Germany. There they are occupied with a certain *philosophy of nature,* of which perhaps only the exaggerations should be blamed in Paris. However that may be, it is not in the judgment of the relation of beings placed at a great distance from one another that this philosophy would have been mistaken.

As this objection is stated, no one I know could take an interest in it. Who ever said that *animals result from one and the same assemblage of organic parts repeating themselves uniformly?* The German philosophy has explained very well that organic parts increase in number and complexity in the sequence of ages, and in the progressions on the zoological scale, according to the order and in direct dependence on the different degrees of organization. We see a simpler organization in the medusa and the starfish, animals that weakness of development have left in the lower degrees of the scale, and, on the contrary, a considerable and complex organization in the giraffe and the elephant, which a more prolonged action of development has brought to the first ranks. Follow this action in a single species, in which the modes of development are marked at intervals by some stopping points. The frog in its mature state enjoys an organization more considerable in number of parts and in vital power than the frog in its tadpole state. It is the same with the tadpole in relation to the egg from which it will come; and finally it is the same for the egg itself becoming cloudy under the influence of the sun, in relation to the egg in its first stage, consisting only of a homogeneous and transparent liquid.

These facts of development by which animals increase in number and complication of parts owe it to a single principle of formation that they are repeated indefinitely in the

formly repeated. Certainly that will not be our colleague,— he is too well informed; he knows animals too well; he knows too well, not only that certain parts are not repeated with uniformity, but that a multitude of parts are not repeated at all.

"In yet another place, he announces that the analogy does not rest on the organs in their totality, but on the materials of which the organs are composed, and he proposes an example, that of the hyoid bone, in the case of which, if we judge by the developments he enters into, he seems to want to give us to understand that it is the number of parts that makes up its principal rule. From some of the phrases that follow, one might conclude that he is adding to this their connections, and in fact, since at the beginning of his memoir he has excluded functions and forms, nothing is left but the connections and the numbers. I do not see a fifth relation, a fifth category, under which one could imagine placing that universal analogy.

"Well! Since for lack of clear propulsion, for lack of a general intelligible rule, I am obliged to lay hold of this theory in the examples given of it, I am focusing on this one. As the vulgar put it, I take our colleague on the very ground on which he has placed himself, and it is thus that I charge myself to take him, whatever other example he may wish to choose.

"I shall therefore examine the hyoid bone in various animals, and I shall demonstrate through the facts, as I have announced that I will always do:

"1. That the hyoid changes in number and in parts, even from a single genus to its neighboring genus.

"2. That it changes connections.

"3. That in whatever way one understands the vague terms used up to the present, analogy, unity of composition, unity of plan, they cannot be applied to this case in a general manner.

"4. That there are animals, a host of animals, that do not have the least appearance of a hyoid, that in consequence there is not even an analogy in its existence.

"Having thus totally annihilated, as far as he is concerned, the principles

---

zoological series. These are the facts that we call analogous, that we say are repeated uniformly, that we seek to lead to generalities, to express philosophically. But certainly no one has had the idea that the medusa was composed, say, of the same twenty-four letters of the alphabet, that these same letters arrived at a certain point, and were repeated to compose the structure of the elephant.

From what suppositions we have to try to defend ourselves!

that are presented as at once new and universal, in whatever sense they are applied, I shall make application of other principles, of those on which zoology continues to rest up to now, and on which it will rest, I hope, for a long time to come, and I shall show:

"1. That in the same class the hyoid bone, although variable in the number of its elements, is nevertheless disposed in the same way in relation to the surrounding parts;

"2. That from one class to another it varies, not only in composition, but in relative disposition.

"3. That the variations of its functions result from these two orders of variations and from their variations of combined forms;

"4. That in passing from the embranchement of vertebrates to other embranchements, it disappears in such a manner as to leave no trace.

"Thus the embranchements differ from one another by the total disappearance of certain organs.

"In every embranchement the classes differ in the connections and composition of organs of the same nature.

"In the same class, families and even genera differ only in the composition and the forms of their organs.

"Here are principles* that at least have the merit of clarity; but above all they have the merit of truth; whatever you may say, it is on them that zoology and comparative anatomy rest. It is in accord with them that there has been formed that great edifice that has been called the system of the animal kingdom.

"And every time that it is attempted to push the generalities further, by whatever name they are decorated, by whatever rhetoric they are supported,

* *Principle* is not a synonym of *result*. With the work of zoology already accomplished, it *results* that animals are finally appreciated in their natural affinities. More simplicity in the number and disposition of the organic parts is the case for some species, and, on the contrary, other animals are the product of the aggregation of a larger number of organs and of a more complex coordination; I add that between the two extremes there are all the degrees of the zoological scale. When this is observed with attention it forms the basis for admirable research and, definitively, for learned classifications that have assisted in the composition of a descriptive catalogue of beings. However, to speak to us of *embranchements, of classes, of families, of genera, and of species*, that is to treat zoology from a point of view that no one contests. What are all these facts doing in the present argument? They are foreign to it. Make sure they are not a veil placed so as to cover the weakness of the reproaches that are being addressed to us.

only persons who know nothing of the facts could take them momentarily at face value, but only to see their illusion shattered as soon as they go on to occupy themselves with investigating the proofs.

"In my following memoirs I shall give the demonstration of this, with respect to every order of organ in particular.

"Today, as I have said, I will restrict myself to the hyoid bone.

"To establish the alleged new principles as applied to it, we would have to be able to maintain that the hyoid bones are composed of the same pieces, that they are in the same connections, and that they exist in all animals."

The Academy will judge whether such assertions will bear the slightest examination.

M. Cuvier will divide his work into two parts: the hyoid bone in animals that breathe air in their natural habitat, and the hyoid bone in animals that breathe only through the medium of water. The latter will demand an introductory discussion of the sternum.

"Everyone knows that, in animals that breathe air, the hyoid bone is an apparatus suspended under the throat that gives attachments to the tongue at the front, supports the larynx in the front, and has the pharynx above it.

"Its name comes from the fact that, in man, its chief part or its body is shaped like the arc of a circle, like the cursive upsilon of the Greeks."

M. Cuvier gives an exact description of this bone, which he examines first in monkeys.

"The body of the hyoid bone in monkeys varies a good deal in form, but that has nothing to do with our discussion. Its posterior horns remain of the same shape and disposition as in man; the anterior horns are generally longer, but also in one piece, and even the ligament that suspends them to the wall is never ossified in any one of its parts, so that even the oldest monkeys never have either the styloid apophysis or the separate bone that serves to replace it in other quadrupeds.

"There we have one difference, but not yet in truth of much importance.

"But here is a greater one.

"In the howler monkey, in which the body of the hyoid bone is inflated, as is known, in the shape of a cucurbit, there is neither any trace of anterior horns, not styloidal ligament, nor anything that recalls the styloid apophysis; the hyoid bone is fixed by other means. How do the unity of composition and analogy contradict themselves so rapidly? Our reply, as to us ordinary naturalists, would be very simple: it is that the hyoid bone, taking on a special destination in the howler monkey, and there becoming a powerful instrument of the voice, needed different attachments: the theory of analogues will not extricate itself so easily from this case. But let us move on."*

*ON THE HYOID OF THE HOWLER MONKEY.

But let us move on . . . On the contrary, I am going to stop at this paragraph, and I invite reflective minds to do likewise with me. The views that divide us show themselves here very plainly: for precise facts, let us give their explanation with rigor.

A long time before the days of our controversy, that is to say, in 1778, the question of the hyoid of the howler monkey was already a matter that had been adjudicated: this was done by the greatest anatomist of that period, the celebrated Camper. With a vast mind, as cultivated as it was reflective, he had a sense so lively and so profound for the analogies of organic systems that he had a propensity for investigating all the extraordinary cases, where he saw not only a subject for problems, but an occasion to exercise his sagacity, which he employed to reduce apparent anomalies to the rule. The publication of the cavernous hyoid of the howler monkey, in the fifteenth volume of the *Histoire naturelle,* had this effect on him and preoccupied him deeply. In 1777, Vicq d'Azir had displayed two howler monkey hyoids at Paris. On his return to Holland, he went through all the rich public and private collections; and after long useless investigation, he finally found in the possession of M. Klokner a howler monkey preserved in fluid, which he obtained and which he brought home, so that he could dissect it without delay.

When his work was finished, he made it the subject matter of a letter that he wrote to Buffon on November 15, 1778.

Camper had been well served in his prediction: he easily reduced all the parts of the howler monkey's hyoid to those of the hyoid of man. "Already," he wrote in 1778, "when I was at Paris with Vicq d'Azir, I had noticed that the bony case, although very thin, was the base of the tongue; I had even distinguished the articulations that had served as the horns of this bone; however, I did not in the least understand its situation and the connection with its neighboring parts."

A cabinet of the Faculty of Sciences of Paris possesses two howler monkey hyoid bones surrounded by their muscles, glands, membranes, cartilages, and laryngeal parts that are attached to them; one of these preparations comes from a male, and the other from a female. M. Hyde de Neuville, who was the minister of marine, had then brought from Cayenne for our cabinet of the Faculty of Sciences. I used these preparations in order to reexamine and understand (with the pieces before my eyes) the drawings and the descriptions that Camper had sent to Buffon— precious materials which had not been printed since 1789, in the supplements, vol. 7. Five figures giving the pieces, one set seen from the front, and the others in profile, leave nothing to be desired, and present a determination such as one would be bound to expect from the fine talent of Camper, that is to say, perfectly exact. All the parts described and illustrated are the same as those of the human hyoid apparatus, with the difference close to their respective volume. The learned Dutch anatomist's views of analogy were fully justified. He saw that the differences of the two analogous organs per-

We cannot follow the author, says M. ***, in all the details that he gives with a marvelous clarity on other species of animals.

After this interruption, the argument is taken up again as follows: "Thus we see that, even in a single class, that of mammals, the number of elements

---

tained to the excessive development of the median part, called the body of the hyoid. In man this median part is hollow, and has the form of a capsule wider than it is high; in the howler monkey, the concavity gains in depth, in such a way that the piece is not very broad, and, on the contrary, extends considerably under the tongue: it is a long, bony pouch, or indeed, as M. Cuvier indicates, a foundation enlarged in the shape of a cucurbit.

M. Cuvier, describing this bone of the howler monkey's tongue in his *Lessons of Comparative Anatomy*, confirms all the investigations and opinions of the celebrated Camper. In the chapter on the hyoid bones in volume 3 of his work, p. 230, my learned colleague has dealt with the hyoid of the howler monkeys only "as presenting an extremely remarkable peculiarity, in that this point serves to explain the howls that these animals produce: the body is as it were swollen to form the bony case. The great horns exist, etc. . . ." However, M. Cuvier, giving a broader path for the spirit of his investigation than had guided Camper at this juncture, dreams of finding some parts that, as he is able to judge, correspond to the anterior horns, which are in fact lacking. *Two small apophyses that rise up from each side of the large opening of the case* are without doubt, according to M. Cuvier, the rudiment of these horns, which would not have been recognized only because they lacked one of the characters of these bones, their detachment from the median piece. I have also just observed these apophyses. I can no longer doubt the correctness of the determination given in 1805; I have reasons for this in other characters that are evident: (1) their being much longer apophyses in the hyoid of females, and (2) the attachment to the stylohyoid ligament and muscle, which are next to the styloid facet of the skull.

Now that I have referred to the work of two celebrated zootomists of our time and have further reviewed and confirmed their findings, I no longer need to insist on that deduction presented above: *how do the unity of composition and analogy contradict themselves so quickly?*

However, there are some remnants of the anterior horns. Thus we find both the ligament and the muscle that accompany it and which together constitute this cord attaching the hyoid apparatus to the sides of the skull. We must also declare inexact that other deduction of the argument, which later is expressed as follows: "We understand that the huge drum formed by the hyoid bone of the howler monkey, fastened by ligaments, and in a nearly immobile fashion, to the lower jaw, did not need so strong an attachment to the skull." We are not unaware that the pieces forming part of the anatomical collection of the Royal Garden furnished a pretext for this statement, but have they been examined with sufficient attention? It was dried preparations that were observed, while I have observed whole pieces, mobile, and perfectly preserved in fluid. From the facts that I have before my eyes, the result is a rigorous determination of the parts that fix the hyoid to the lower jaw. I affirm that they are not of the nature of ligaments: I guarantee that they are muscles, and precisely the muscles that analogy would have inspired me to look for in their usual places. Thus it is at the front, the genio-hyoid, that Camper designated by the letters A to

of a single organ, of the hyoid, has nothing in it that is constant; there are what I call variations of classes, that is to say, differences in number and much larger differences in form, but still an almost absolute resemblance of connections.

G in the figures published in Buffon's supplements. (See *Hist. nat. générale et particulière*, supp. 7, pl. 27, fig. 1.) At the sides are the mylohyoids. Camper also represented perfectly the muscle that is decisive for the question that has been raised here, to wit: the stylohyoid. (See a B, fig. 3.)

All these facts are differently presented by M. Cuvier; I am obliged to say of some that they are reported inexactly. Thus, it becomes useless to debate an explanation that is their consequence. Otherwise, if it was necessary to seek in that explanation everything that it contains of value and of valid conclusions, I would have to reproduce the claims that I made in the note above. Yes, without doubt, it is not philosophical to explain the production of a new organic device through new habits, or to satisfy *a special purpose*. And in this species we have a decisive proof of this; it was said that it is because *the hyoid of the howler monkey becomes a powerful instrument for the voice that it needed different attachments.* We have just seen that these supposed new attachments constitute an inexact fact.

It is at this moment that the argument claims to have finished with the hyoid of the monkeys, with these words: *the theory of analogues won't get out of this so easily!* I cannot keep myself from remarking that this moment is badly chosen. There are no ligaments making attachments, nor is it necessary that there be ligaments to attach the hyoid body to the lower jaw.

But up to the present we have employed only observations and inferences as the Aristotelian doctrine and the methods improved by recent anatomists would suggest; let us arrange for the theory of analogues, which until this moment has figured in this note only insofar as it has been attacked, usefully intervene here to some purpose.

The two principal differences of the howler monkey's hyoid, compared to the hyoid of man, on which the works of 1778 and 1805 have not placed sufficient emphasis, are these: (1) the very considerable volume of the body of the apparatus, and (2) the absence of the anterior horns, or at least the fact of their articulation by synarthrosis.

On the first point the reply is simple: the volume of the parts becomes a very important circumstance in each species separately, for in each case it regulates the function by procuring for the organs all the power they can acquire; but this is a consideration that philosophical studies neglect, and must neglect.

On the second point, the theory of analogues could not remain satisfied with the remark, otherwise judicious, included in the *Lessons of Comparative Anatomy;* it is not enough to admit as a certain fact that the articulation of the same horn is established in man by diarthrosis and is transformed into an articulation by synarthrosis because of the joining of that same horn to the median body. Here is why: it is because man himself, with respect to his hyoid organ, does not meet the general criteria for the class of mammals. But the theory of analogues demands such criteria everywhere: thus, if the normal number of parts is different, the theory of analogues cannot fail to assign the causes for that difference.

"But if we pass to the class of birds, that is something else again; a great and perceptible hiatus!*

"No further suspension to the temporal bone; no more posterior horn; a body directed horizontally, terminating in front in an elongated growth, a kind of tail, on which the larynx rests, and which often forms a bone on its own; only two horns, each made of two pieces, articulated underneath, at the side of the body, at the spot where it is itself articulated with its tail, turning around the oc-

---

In the howler monkey, in the spider monkey, and even in the Old World apes with the hideous faces, known under the name of baboon, the styloid chain consists only of a ligament, while in the four-footed mammals, it is formed of three ossicles in transversal series.

If the theory is mistaken in its prediction as to the number of pieces, it has recourse to another of its rules, with a result that is different but no less effective for a second prediction: it admits that one of the pieces will be nourished at the expense of its neighbor. This rule, known under the name of *balancement* (between the volume) *of the organs*, explains the hypertrophy of one of the materials through the atrophy of one or several others.

What could have provided for the enormous increase of the hyoid body? Necessarily, a sacrifice imposed on the neighboring pieces. But those which their situation calls to support all the effects of sacrifice are necessarily all the ossicles forming part of the styloid chain: when these chains are struck with atrophy to the zero point of their bony molecules, nothing is left but their periosteum of cellular tissue in the form of a ligament.

Thus what the theory of analogues does not encounter in number of parts, according to the prediction it is inclined to following the table of its observations in the majority of animals, it finds in justifications, in *compensations* that it knows how to discern, in rudiments that declare the why and the how of the disappearance of certain materials.

*It is not at all on this terrain that I challenge the efforts of the argument. It is indeed true that there is a hiatus here, that is to say, that there exists a hyoid truly special to the class of birds: but this fact is not repeated here except after I had established in my *Anatomical Philosophy*. This is not the moment to add that I do not believe I have ever produced anything more directly useful to the theory of analogues than my particular writing on this matter. Before my research, it was held that the tongue of birds was bony, or at least that, to furnish a support for it, as the chest finds one in the vertebral column, in the birds ossicles intervened for the benefit of the tongue, suddenly and extraordinarily, ossicles of which there was no analogue in mammals.

I have been so prolix in the preceding note, and I have so much to add to my earlier writings, since I have to extend them in order to correct some errors, that I am obliged to stop these reflections here. But this is to establish in a memoir ex professo all the facts and all the corrections that I have accumulated for several years pertaining to the first youthful works. This memoir will appear in the sequel that will follow the publication of this pamphlet.

ciput, even going, in the green woodpecker, as far as the base of the beak; and at the front the body carries a bone, or two bones attached to the sides of the other, articulated at the front end of this body, and forming the skeleton on the tongue; for the tongue of birds has a bony skeleton of which there is no trace in mammals.

"For everyday vision, for the appearance as ordinary common sense grasps it, there was nothing to reply; here is a great change of composition, and just as considerable a change of connection. We see that we have passed from one class to another.

"What does our learned colleague do for lack of an argument?

"He has assumed that the hyoid bone of birds, derived in one part from the muscles of the tongue, and in the other from the larynx, have undergone a rotation on their anterior horns, and that their posterior horns have thus found themselves directed to the front and have become the bones of the tongue.

"Here no doubt we have a somersault possible to conceive of in skeleton where the bones are held together only by a brass wire, and where there are only bones. But I ask any one who has the slightest idea of anatomy: is that admissible when we think of all the muscles, of all the bones, of all the nerves, of all the vessels that are attached to the hyoid bone? We would need to—but I stop! The very idea would frighten the imagination. To preserve an apparent identity in the number of bony pieces, one would have had to change everything in the connections and in the soft parts. Then what would have become of the unity of plan? But let us not make any prejudgment, let us admit for a moment so strange a hypothesis; let us see if it will lead us much further."

(M. Cuvier passes to a third class, to the reptiles,* and taking the tortoise as an example, he refutes, following the same procedure, every idea of an analogy between the hyoid of that animal and that of mammals and birds. Then he adds:) "People who admit a degradation, an insensible simplification of beings, a principle, to mention this in passing, absolutely contrary to that of the unity of composition, which is nevertheless allied to it in certain minds, so great are the extravagances in some heads—those who admit such a degradation are going to suppose that other saurians have hyoids just as simple as or simpler than that of the crocodile. There is nothing in this.

"In lizards with a protractile tongue, the hyoid bone is more complicated

*The reptiles do not form a *natural* class, especially in the way that birds do. I have always wanted to explain myself in this connection, and I anticipate writing on this subject, which will require very extensive developments.

in its forms, more singularly refolded in its different parts than in any of the preceding animals.

"All these facts are incontestable; any one can assure himself of this at any moment; by what effort of reasoning will we be made to believe that there is identity of elements, uniform repetition, identity of connections, indeed, all those expressions that are employed one after another, between hyoid bones of which some have only two pieces, others only four, while there are some that have seven, others nine, and even more? In the soft-shelled turtle up to seventeen or more can be counted. By what art will they succeed in convincing us that there is identity of connection between hyoid bones some of which are suspended from a part of the temporal bone, while others surround the skull and penetrate as far as the beak, and while yet others remain absolutely recumbent under the throat and as it were drowned in the muscles? Is there anything seen there that we have not been seeing for centuries: a certain resemblance, the degree of which is proportionate to the relations of animals to one another, and differences determined by the use that nature makes of this organ, or, if one wants to avoid every shadow of reference to final causes, differences that determine that use?

"For us ordinary naturalists, these relations, these functions, these differences, are explained very well, since they make the animal what it is, since they invoke or exclude one another.

"We understand that the huge drum formed by the hyoid bone of the howler monkey, fastened by ligaments and in almost immobile manner to the lower jaw, had no need of so strong an attachment to the skull.*

"We understand that the nearly inflexible tongue of birds had to be carried at the front by a mechanism different from that of quadrupeds, which can be contracted in all directions; that since their larynx has no thyroid cartilage, the posterior horns of the hyoid can be lacking; but we do not understand how, by a rocking motion that would have destroyed all the muscles and all the vessels, they would have gone to settle in the tongue, etc.

"But if we neglect all these considerations, to see only alleged identities, alleged analogues, which, if they had the least reality, would reduce nature to a kind of bondage in which, happily, its author is far from having enchained it, we no longer understand anything about beings, either in themselves or in their relationships; the world itself becomes an undecipherable enigma.

"I know very well that it is easier for a student of natural history to believe

*I have included the developments of this paragraph in the long preceding note.

that everything is one;* that everything is analogous, that through one being one can know all the others, as it is easier for a student of medicine to believe that all diseases amount to only one or two; I even admit that the error that the former would induce would not be as fatal as the latter, but it would still be an error. We would be casting before our eyes a veil that would conceal true nature, and the duty of scholars is, on the contrary, to push aside that obstacle to the knowledge of the truth.

"In the second part of this memoir, which I shall forthwith have the honor to read to the Academy, I will treat the hyoid bone in frogs, in salamanders, and in fishes, and I will show that it is by transpositions and see-sawings even stranger than those of birds, that it has been supposed one could find there identities of numbers, which, even admitting all the presuppositions, would not be there at all.

"Then I will show that the hyoid bone is lacking absolutely in an immense throng of animals, so that, whatever meaning is given to the theory of analogues, it is impossible in its case to make a general application.

"I repeat that it is with much displeasure that I have seen myself constrained to break a silence on which I had firmly resolved, if I had not been pushed to the wall; but in fact naturalists would have the right to blame me if I abandoned so clear a cause.

"What it is above all essential to repeat is that it is neither to confine myself to old ideas nor to repel new ones that I have made this defense. No one thinks more firmly than I do that there is an infinity of discoveries yet to be made in natural history. I have had the good fortune of making some, and I have reported a great many made by others; but what I also think is that if anything could stop us from making true discoveries in the future, it would be to want to confine minds within the narrow limits of a theory that is true only insofar as it is old, and which contains nothing new except the mistaken extension that is attributed to it."

After the reading of this memoir, which excited the interest of the Academy in the highest degree, the floor was given to M. Geoffroy Saint-Hilaire. That learned naturalist read the second part of the memoir in which he develops his *theory of analogues*. We regret that we cannot reproduce it today; we will soon make amends to our readers for this omission.†

* The preliminary discourse has answered this part of the argument.

† This promise was not kept. NO extract was given either of this memoir or of my lecture of March 29; but on the occasion of the latter, and in the summary of another session

N.B. M. Cuvier's third argument, dealing like the second with the modifications of the hyoid, is from April 5, 1830. I am not replying to it in this pamphlet; I shall do so in the next installment.

In replying on the 29th to the text of March 22, I distinguished the general question from the particular facts; I had already dealt with the former, further on, in my *first reply,* when I became aware that reasons of moral propriety (see the annexed exordium) demanded the interruption of our discussions by verbal pleadings. Thus it remains for me to deal with particular facts; and it has already become clear in a preceding note that the classic hyoid of the birds alone will form the material of a separate memoir.

---

of the Academy, use was made of the *System of Differences,* since nothing was seen in my works except *overabstract considerations,* and repugnance was shown for a *philosophical principle in which one had to believe as if through one's emotions, like a revealed truth.* Other public documents treated me with more favor.

ON THE HYOID BONES

First reply to the previous argument

(SESSION OF MARCH 29)

I believe it due to the dignity of the sciences to preserve a tone of decency with respect to persons and manners, showing esteem and good will. Since, in extending my research so far, I have been exposed to the danger of erring, I am indulgent to an error conceived and produced in good faith; for even fruitless efforts always remain estimable, and as it were an indirect homage to the truth and a testimony to zeal and devotion. I also believe one ought to avoid transforming a reunion of disciples of the Stoa into a crowd in the pit clapping their hands at the outrageous comedies of Aristophanes. Before the serious public that is listening to me, and since I am dealing with serious matters, I shall be sober and never witty. Consequently, I shall prevent myself from hastening the moment when this truth can come to light and appear in all its glory—which will happen only when it is incontestably established.

I have encountered some alleged conciliators who boast of having penetrated the secret of our disagreement; to hear them, they are going to teach us this point, of which we are ignorant, and reconcile us, "for, in fact," they say, "each one is following a particular path, this one, when he pursues the facts with respect to the nature of their differences, and that one, with respect to the nature of their relationships; that is to act for the best from two sides, if, on the two sides, each one remains equally faithful to his point of departure."

Unfortunately, I can admit neither this conciliation nor this reasoning: I have no faith in an exploration of facts; I trust to a profound knowledge of things only as long as investigations have been exhausted on the differences as well as the relationships. Neglecting one aspect of one's subject so as to pay all one's attention to the other is the way to know it only imperfectly. If, however, we cannot separate the study of the relationships from that of the differences, and vice versa, the whole problem of the determination of the organs depends on the choice of a method that will arrange and coordinate the facts, just as well for one point of view as for the other.

I am asked to be more at my ease; where I believe this to be useful, I shall be so. Caught up in a European movement, I second it as best I can. As far as possible, the old paths of zootomy are being abandoned; anatomists are seeking new paths; let us act so that France will not lag behind.

But I cut short that preface to come decisively to the facts of the argument of March 22. So many details on the hyoids are imposing: the public is bound to let itself be caught as by the evidence of vast knowledge; as numerous as the facts are, I do not fear them, and I would even take them willingly to be very accurate, if it were not, nevertheless, for that wonderful brass wire, capable of executing, in a skeleton, such an ingenious maneuver. A jest is not to be commented on; I pass it over.

Admittedly, it is appropriate for an honest discussion to have arrived at the point of studying the general question in a particular application; and the choice of the hyoid is especially happy if this is to be done in depth. In this respect, the facts observed are so obvious that, as has been remarked, and I am of the same opinion, only common eyes are needed to see them, only ordinary common sense to grasp them; consequently they are, at least most of them, such as the argument has arranged and presented them. That given and agreed to, it will be asked if after that admission there still remains a disagreement between us on the character of the hyoids; yes, without the slightest doubt. For it is a scientific appreciation of these same facts that is the matter at hand. It is a question of philosophy that divides us, not, however, in such a high degree as some seem to believe and as has been said. To keep us at a distance, there is only the interval that separates the ideas of the Aristotelian doctrine from those of the theory of analogues. That is what has to be explained.

It is not without mature reflection on this matter that I have just rejected the path of conciliation that has been offered. The proposition was equally offensive to both of us: for if one of us is not excluding the relationships in order to attach himself solely to the consideration of differences, no more does the other mean to neglect the differences in order to occupy himself solely with the relationships. Must only the differences be studied? There is great merit to the use of our senses at certain materials, if it is only a question of counting or at organs whose weight or length one wants to measure. We know some naturalists, qualify them as you wish, who confine themselves to such superficial work, and they are not to be disdained. And among these works I cannot, nor do I intend to, include these *Lessons of Comparative Anatomy.* Certainly, I take too much to heart the observations of the conventions, the desire to be fair, to allow me to do this, even by an indirect allusion.

However, our views do differ. In what does this difference, whether of method or of philosophy, consist? If it happens, in a given case, that attention is turned with preference to the character of the differences, the relations are unfortunately admitted ahead of time and not after a study ex professo. You have a presentiment of the relations, you at least hold them as acquired instinctively. In some cases, but not always, you have the evidence on your own. One is indeed authorized to say, and entirely excused from proving, that, for example, the bovine eye is in all respects an organ identical in composition to the human eye; just as, in the science of numbers, we declare, and do not prove, that two and two are four. But I repeat, it is not always in this character of isolation, not always with so evident a revelation of their common relationships, that the comparable organ systems of animal organization present themselves. It is quite simple, if an instinctive conception persuades you that human and bovine eyes are at bottom one and the same organ, that you could afterward pass to the comparison of all the details, and that you could examine all the differences between them. Every part can be more or less emaciated, more or less voluminous, and the sum of all these partial differences gives the differential characteristic expression of every eye in particular.

But for one such simple case, how many others are there that offer a very great complication, and that constitute curious problems to be disentangled? Since I have written on the hyoid ex professo, the fact is that I have considered it to be placed in that second condition.

The argument to which I am replying seems to me to rest on a continued contradiction. It calls the hyoid of man different from that of the ape, that of the ape different from that of the maki, different from that of the lion, etc. But what student of zoology does not know that? When we pass from one class to another, the hyoid, by the same argument according to which animals descend to some degrees, is modified more profoundly. What are we to conclude from this exposition? Doubtless nothing, except that these facts are perfectly well known. Much emphasis is laid on the hyoid of the howler monkey, on that enormous drum in the shape of a cucurbit. I had expected to be taught something new on this subject: whether this is a bone hollowed out like the cranial cavity. Has it been examined in its earliest stage, to know if it is equally formed of parts? What they wanted to say is that there is a difficulty for everyone. I do not at all agree with that, and I return to what I have written in my sixteenth lecture on mammals concerning the cavernous hyoid of the howler monkey.*

However, by enlarging the differences to make them stand out from cases of proportional alterations in the degree of organization of each family, would the argument have seemed to be saying that the differences are so strong that, by themselves, they dominate, and that the relationships are nowhere? Isn't that what we would have concluded from the phrase "*turn everyday eyes to the facts, they can only stick to the appearances; they see that there is no resemblance*"? But I say in my turn: "turn to the facts the spirit of combination and of investigation, come to them with a sagacity capable of grasping the common points, hidden under the mask of some excesses in the volume of the parts, concealed by forms that cases of hypertrophy or atrophy had profoundly altered, and you will soon perceive the analogy of these facts; just as certainly you will easily assign their relations."

There is no doubt that I will do no harm to the argument if I say that in fact it fails to recognize the general propositions that result from these relations; for it would answer me: *Don't I give the same generic name to the hyoids everywhere, as much to the mechanism so designated in man as to all the families of the four classes of vertebrates; and to give the same name to a thing: is that not to declare implicitly that one believes in its character of being at bottom one and the same thing?*

Thus I have to be the one to take pains to discover in the inferences of the argument that it is definitely aligned with my view, and consequently that it and I both believe in a hyoid, the same from a philosophical perspective. But then I can reply: why give oneself so much trouble somehow to conceal this truth, to bury it under so considerable a heap of differential causes, all quite good to point out, if one restricts oneself to their bearing as particular facts?

There is at least contradiction in the inferences of the argument, if it maintains that there is no hyoid essentially the same with respect to its intimate composition, when it uses the same word to designate it. For I do indeed think that the argument would have wanted to save the trouble of teaching me—a person

*Such was my first reply, already made according to my recollections; it is given here as I communicated it to the Academy. However, in fulfilling the duty of correcting the proofs as to the previous subject, I have been able to take up the same matter again, to review it by interrogating the facts again and to conclude with an extended discussion on this particular point of our controversy. That is how the question relative to the hyoid of the howler monkey came to be reproduced for the second time in this work, and is even already treated above.

who has written ex professo on hyoids—that in examining them in their secondary modifications, there is really a diversity of the hyoid from one family to another, and a greater diversity from one class to another. Otherwise I would also enter into the details, and I would show that after studying that structure first in its relationships, I had arrived at a deeper knowledge of the differences: I would show above all that I had not confined myself to naming the weight and the length of every part, which forms the most considerable portion of every description of facts, but that I explain why and how the differences arise. For if one element is absent, the principle of balancement of the organs gives the explanation of that fact: that is to say, that this absence shows itself at once compensated for and revealed by the growth assumed by another organ in the vicinity. The fact is that there are only well-understood differences, only the differences ascertained by a complete exploration of the facts that would have specified their mutual relations in advance.

We cannot be silent in this regard; there is a manifest confusion with the reasonings of the argument; and this confusion seems to be carried to its extreme when, taking no care to discern the various degrees of organization, the argument asked that it be furnished ipso facto with the immediate relationships *of the Medusa and the Giraffe, of the Elephant and the Starfish*. Doubtless such an a fortiori does not arrive at that point except *in despair of an argument*. I truly regret recalling that expression. This a fortiori is probably no more than a slip escaped from the pen of my learned colleague. I believe the phrase brought a smile to someone in the audience; but I suppose, at least, that it would not have brought conviction to any mind.

I attribute the confusion I have just been pointing out to the difference of our two methods. The Aristotelian doctrine, even as it has been recently improved, still abandons much too arbitrarily the point of its departure in the investigation of analogous organs. It is satisfied if between the organs that it considers comparable there is some relationship of form or function that strikes the senses. It seizes on that relationship without any other justification. Thus its fate, or I could say instead its fault, is to admit the fact of analogy before study, to pass at once to accessory modifications with the character of differential causes. And in practice, I have so often paid attention to studying its resources, and have always seen that its practitioners are carried by this doctrine even beyond the goal they have set themselves. In fact, when you make your way in the ranks of different families, and if, as you descend some degrees in the scale, the differences increase in intensity, you neglect to verify whether serious and proportional changes had not intervened on the primitive conditions of the analogical fact.

The theory of analogues, on the other hand, defends itself against this vagueness; it prevents all confusions by the severity of its point of departure. If an apparatus is composed of several materials, the theory is satisfied only if it knows each in its essence; in attending to the differences, it never loses sight of its point of departure; it notices if materials disappear either by soldering, because there was a fusion of one piece with another, or by an atrophy carried to its extreme. For the theory of analogues does not presume the invariable conservation of the materials, but intervenes to appeal to them and to determine their number. Thus, it is after a study ex professo of materials that, departing from the relationships it had studied in advance, it frees every faculty for the considerations of differences.

It is neither these principles, nor any of the corollaries of my work ex professo on the hyoids, that the argument recalls; but it is conceived from prejudices that it then quite easily combats. "*Your chief rule*," it states against me, "*recognizes only the number of parts.*" However, that is not true. You may judge of this by the two following corollaries from my memoir published in 1818:

1. The hyoid apparatus is basically the same in all vertebrates.

2. The hyoid, generally speaking, is composed of nine pieces in the fishes, of eight in birds, of seven in mammals, not counting the styloid bones.

However, the argument adds that from certain phrases in my later memoir, it can again conclude that I am also relying on the order of connections, for these phrases clearly announce the exclusion of consideration of form and of functions. I reply, likewise, with another corollary of my work of 1818, where the fact of these relations of connection is posed as a principal character: on every occasion, I wrote, the hyoid forms *the solid framework of a partition that separates the back of the mouth of the vestibule from the respiratory organ.*

And continuing with the charge of the lack of a clear statement, of a lack of a general intelligible rule, the argument claims to prove by the facts:

1. *That the hyoid changes in number of parts from one kind to another.* I have said, I have established, I have already proved that. Every class, not including that of reptiles, which is artificially formed, sees recurring for itself a given number of materials, nine, eight, or seven: if that does not always hold with respect to some forms, the exception goes to confirm the rule. For the perturbing cause is then shown with clarity and gives reason to the apparent disorder.

2. *That the hyoid changes connections.* That is what the argument announces, and I promise not to abandon this terrain myself; I shall explain myself more clearly in a moment.

3. *That in whatever manner* (I am transcribing), *that in whatever manner one*

understands the vague terms of analogy, of unity of composition, of unity of plan, employed up to now, one cannot apply them in a general manner to the hyoid.

I have replied above to that assertion, and I have, I believe, sufficiently demonstrated that, combined with the use of the word hyoid, this objection is meaningless. And in fact the idea of the generality of a hyoid apparatus, which is basically the same for all vertebrates, is rejected precisely in a dissertation in which that entity is given a general name. Whatever one may say, this is an organ sui generis, and certainly the hyoid has to exist before the faculties that will ultimately be acknowledged in it, that disposition of parts that they would like to make the exclusive subject of the considerations to be applied to it.

4. *And finally, that there are animals, a host of animals, that have not the least appearance of a hyoid bone and that consequently there is not even an analogy in its existence.*

I cannot believe that it is for me, for a scholar versed in zootomic studies, that that objection was written. There has to be the right time, the suitable age, for the hyoid to appear in any embryo whatsoever, of man, of mammal, of bird, etc.; before that it is not compatible with the degree of organization of that epoch. Even in animals that belong to the same degree of organic development, the hyoid can be absent; what is surprising about that?

Shall I add some reflections on the form of all bony tissues? I shall not be exposing myself to ridicule by seeming to teach something on this point of theory to my learned colleague. And in fact, whom is it necessary to persuade that the hyoid, like all other bony tissues—that the hyoid, I say, before having the character and consistency of bone, has passed through the cartilaginous state, that before that it was in the fibrous state, and that, still longer ago, it was represented by an aponeurotic membrane?

I had refused to believe that it could have been brought as an objection against me and as a new proposition that the materials of the hyoid disappear, that there is no hyoid in animals of the degree of development that characterizes the organs of embryonic life. Have I not written, on the very subject of the hyoids, an organ is destroyed, and vanished entirely, rather than being transposed?

Now the argument continues: *I have destroyed, I have totally destroyed the principles that have been given at once as new and universal: nothing remains for me but to apply other principles on which zoology has rested up to the present, and on which it will long continue to rest.*

This ancient basis for zoology consists in the consideration of forms and functions; that is what the argument will try to take up again, but by making a great backward step. There is great shrewdness in the choice of a word that is used for the first time; for through its double sense, the attempt is made to place oneself in the middle of the distance that separates the two doctrines, that of Aristotle and the theory of analogues: that is the word *disposition,* which is certainly a clever invention; for according to circumstances, it will lend itself to signifying *position of parts,* in the study of anatomy, or *relationship of functions,* in the study of physiology.

A moment ago it was four objections that annihilated, totally annihilated my principles. Here come four propositions corresponding numerically to those, which will contain true principles in replacement of the false ones. Here they are literally; reflections on them will follow.

"1. The hyoid bone, in a given class, although variable in the number of its elements, is nevertheless *disposed* in the same way, in relation to the surrounding parts.

"2. From one class to another, it varies, not only in composition, but in relative dispositions.

"3. From their two orders of variations and from their variations of combined forms result the variations of their functions.

"4. And finally, if we pass from the embranchement of vertebrates to other embranchements, it disappears so as to leave no trace."

I, I who understand the sense of these words, I see with pleasure that I no longer have any adversary in what concerns the hyoids, as far as generalities are concerned; perhaps on one single point, the chapter of connections.

I said above that the choice of the hyoid in the present discussion was happy, because I already had a presentiment of the actual result. The number of pieces is limited, and as well known as they are to both of us, they ought to speak with authority to one of us and to the other. In short, another order of reasoning ought to guide the conciliation of the two opinions: it is that under the action of the one as of the other inspiration, pressing the lines of reasoning to each point of departure, and finally considering the pieces, whether anatomically or physiologically, one could only come to judge them in the same way.

What is contained in the four questions, or the new principles to be substituted for those announced by me?

1. It is admitted that *the hyoid is composed of a certain number of elements and that it is* DISPOSED *in the same manner in relation to the surrounding parts.* Except for the new use of the word *disposed,* but which in this phrase is certainly synonymous with the adjectives *situated, placed,* that is one of my corollaries. Before me, who would have thought that there was something like an apparatus

in the hyoid bone, and that that apparatus was composed of elements each separately determinable?

2. How is one to understand *that from one class, the variation no longer carries an understanding of the composition, but only of the relative dispositions*? If I try to grasp the rather obscure meaning of that phrase, I believe that *relative disposition* is there to take the place of the expression *relative function*. But it is not I who would object to the trouble of an investigation concerning the function; I ask only that it take place consequently to the determination of the hyoid body, or better, of the various hyoid elements.

3. *The variations of the function are the results of other causes of variation*. I adopt without difficulty that general proposition, which agrees with the sequence of my ideas. It is a long while that, objecting to the teaching of final causes, I have been saying: *Such as the organ is, such will be its function.*

4. *Finally, if we pass from the embranchement of the vertebrates to other embranchements, the hyoid disappears.* I believe I have observed that it still persists in the crustaceans; but let us pass over that: there is nothing then to which I have not already replied completely, categorically.

Thus the theme I had produced in my memoir ex professo on the hyoid has been remade. Without letting this appear too obviously, a few steps backward have been taken in approaching me; but it must be true, that is not yet the whole distance that has separated us.

The mistake made, according to me, is to take the hyoid as an abstract entity, before the study of its relationships, then to develop all its differential faces, when, on the contrary, I do not move to those cases of difference before having reproduced all the elements of the hyoid apparatus to their true analogues. Given that these elements vary in number following the families or the species, I wish, before comparing them, to know what it is that I am to compare: I ask how many materials are used in the formation and the composition of the hyoid: how many, and what ones in particular are conserved to form part of the apparatus?

Now, in other respects, it is not for cleverness, that I shall praise the argument. It could have found where to catch me out, if it had discussed the applications I had made of the principle of connections; the argument did not do this felicitously: it produced general allegations, but no positive explanation based on a demonstration. The trouble is that to do better, they would have had to attach to the character of connections as much importance as I do: which is not the case.

I shall have to review some early works: errors were inevitable in an enterprise continued for so many years. These mistakes are reparable, and have almost all been effaced, on the very indications of the principle of connections; that is to say that I must not deviate from it in any way.

When I took the courageous decision to achieve ex professo the determination of every organic system, everything was yet to be investigated, to be created: both principles and method of experimentation; but above all it was necessary to protect oneself against vicious habits that did not allow further progress. Had I to arrive at all the unknowns of all the problems at once so that I could rely on some for the benefit of the others? This is what had been done and to no good purpose. I, on the contrary, took the position of concerning myself only with a single system, with trying to compare it in isolation and part for part in the whole series of beings. I chose the bony system. Once this unknown was disengaged, the other unknowns, or so I hoped at least—that is to say, the other organic systems, nervous systems, circulatory, muscular, etc.—could not fail to be illuminated in a vivid light by the facts studied in the isolated unknown or in the organ that had been determined.

However, in order to provide all the desirable generalities, to pronounce with equal certainty on all the difficulties of the subject, the science of organization did not yet have at its disposal other resources, on which that of the determination of the organs has since found it can rely. O! if that assistance had come to us from abroad, from Germany for example, which so much work in this direction makes so advantageous to the friends of the sciences, with what energy and enthusiasm we would have expressed our gratitude! What satisfaction we would feel in celebrating such great successes! But this obligation, which we have to one of our own, an anatomist placed in our ranks,* and another sentiment that is also the accomplishment of a duty, oblige us to speak with reserve of the extraordinary assistance which, in recent times, the doctrine of the unity of composition has received from the theory of eccentric development.

Another unexpected resource, which also came equally to assure my progress, has been furnished to me by my studies on monstrosities. All the facts of variation that the series of beings had offered me, monstrosities gave me again in a correspondence that followed and was in some sort regular, by means of the anomalies that are repeated under diverse forms in the circle of developments of a given species.

This is why and how I find I must make some corrections, relative to my

*Dr. Serres.

former determinations of the materials of the hyoid period. To announce these corrections is to promise a new work: I reserve it for another session.

I have not had time to follow the argument of March 22 on the hyoid bones in all its amplitude, that is to say to take up again all the details that it accumulated. It seemed to me that it was necessary to treat the general facts first: on a second level, the details are no more than consequent facts, which it then will become very easy to arrange each in its place and to appreciate exactly in its particularity.

Various Concluding Reflections

But there is a better point: it is that if I had been mistaken on all the facts of detail, the argument would yet in no way have the right to reach a conclusion against the principle of my philosophical doctrines. For it would not be the first time that a generalization would be considered legitimately entered into the treasury of conceptions of the human mind, although at first it had been based on some particular inexact considerations, on proofs now held to be inadmissible.

Thus Buffon erects into a zoological law the proposition that the animals of equatorial countries inhabit one of the continents to the exclusion of the other; Lavoisier offers his theory of vinous fermentation; and Lamarck, with the same certainty of mind and judgment, claims that there are in the external world causes of influence and of excitation sufficient to modify the organization of animals by means of their action, sufficient to alter their form, to make their functions vary. But these propositions, conceived with the absolute power of intelligence and foresight, today universally authenticated, were nevertheless supported at their first appearance only by demonstrations founded on facts of which the experience of later years has revealed the inaccuracy. However, it will be said, how could these general propositions be true, and then false the facts from which they had been deduced? That is because there existed beyond the observed facts something more for the thought of those men of genius. Such were in effect the right and the due of their superiority of intelligence, that they held to be truly existing what, in their force of conception, they had judged must be. Thus, for their lofty abilities, since the facts were necessary, they were anticipated, preperceived, inferred.*

*Do you want an example of this absolute power of genius? Hear Montaigne when he has been describing a monstrous infant, of the class of heterodelphs. Montaigne, who always relies on the rationality of things, knows, but rejects, the ancients' explanations of

The argument passed over in silence the lofty themes of philosophy. Nevertheless I am thankful that, in proposing to invert my doctrine, it dreamt of replacing it by another order of the linkage of causes and effects. "Such is its principle of the conditions of existence, of the conformity of parts, of their coordination for the role that the animal HAS to play in nature."

However, that is to substitute for considerations of *fact* considerations of *needs*. That is, by whatever means it is attempted to conceal this nevertheless manifest intention, to resolve to attach oneself to the facile and deceptive explanations of final causes. I shall not return to what I said above concerning this philosophy; I can only believe it to be generally abandoned, when I read these words, which appear to me of a profundity and a force of truth to be at once grasped by all reflective minds: "Despite their name, *final causes* are nothing but obvious effects, or the very conditions of existence of each object; and in this context, it would perhaps be better to call them necessary causes. In any case, it is certain that nobody has ever proved anything by them, unless it is their very impotence to prove anything" (*Revue encyclopédique*, vol. 5, p. 231).

These last reflections on the connection of facts, on their necessary causes, seem to be attached only indirectly to the questions debated in the present controversy, but their common context could not escape the sagacity of the reader.

Let the reader in fact have confidence in the advances of public thought; let him be a man of his time, let him use his faculty of judgment, and let him not be prejudiced by that principle so often intentionally reproduced—that *natu-*

---

monstrosity. Aristotle saw in it only a subject for the condemnation of nature as abrogating her laws; and Pliny, reviving that thought by an exaggeration, had said: SHE WANTS TO ASTONISH US AND AMUSE HERSELF; *miracula nobis, ludibria sibi facit natura*. From the single fact that he has before his eyes, Montaigne rises to all the heights of the question; he judges the phenomena of monstrosity according to their causes and necessary conditions, and he concludes as follows: *what we call monstrosities are not that to God, Who sees in the immensity of His work the infinity of forms that He has included in it.*

This thought of Montaigne's will be developed. Hérboldt, the celebrated physician of Copenhagen, already considers monstrosity as permanent cases of pathological anatomy, as a fertile source of instruction, showing various other possible arrangements as to the circulation of fluids.

There is no doubt that the facts of monstrosity, collected and analyzed, are becoming for the study of animal organization a sort of separate science of the greatest utility. Its *Elements*, where the known facts are conveniently collected, is a book that is necessary today: my son (ISID. G.S.-H.) is employed in writing that work: he has opened it with a thesis that has brought him to the attention of physiologists, by his inaugural medical thesis entitled *Propositions on Monstrosity, Considered in Man and Beasts.*

ral history is the science of particular facts—by the spread of the view that there is no philosophy except with numerous facts, cunningly disposed, except by them and with them, legitimated and recommended as they are under the name of positive facts.*

For is it the case that anyone would know facts to which this qualification does not at all apply? Would you like to insist that naturalists do not recognize its necessity? We must not press too far this point of the argument: that would be to embarrass too much. The accusation falls or initiates the work of a very grave accusation.

But, however, there is a certain school that abuses the a priori method, which imagination carries to the degree of poetry and which, formed chiefly by the *philosophers of nature*, develops from its confidence in its presentiments a means of explanation for the solution of the highest and most difficult questions of physics. But let us, we say in our turn, also think of that other school, which is too eager that we stick exclusively to the mere establishment of facts. Or

*M. Cuvier relies heavily on the strength of influence of that expression, and opposes to it the tendency of some minds, which he views as unfortunate. Thus when, on October 12, 1829, he presented to the Royal Academy of Sciences the natural history of a new genus of parasitic worms, *Hectocotylus octopedis* (a memoir that was published in the *Annales des sciences naturelles*, vol. 18, p. 149), he did not fail to insist on the remark that, to explain this novelty, another person in his place would have hastened to build a system: such were his words, which the elevation of his voice and the pointing of a glance carried in my direction: *As for us, who have long professed to hold ourselves to the exposition of the* positive facts, *we shall limit ourselves to describing.*

At the following session, on the nineteenth of the same month, I replied to his insinuation; that was in my essay *on the two Siamese brothers, attached belly to belly since their birth.* After presenting in this memoir my views on the law of the formation of organs, I continued in these terms:

"Yet it is not a vain product of the imagination, but a point of achievement of scientific destinies and duties, which arrive at their due times, born as they are of the advance of the human mind; so that nobody should misapprehend the meaning of these words, we shall add that after the establishment of the *positive facts*, the scientific consequences must follow; just as after the work of stonecutting, the use of the stones must follow. Otherwise, what benefit is to be derived from these materials? True disappointment if they are unused, if nobody assembles them and uses them in an edifice.

"Like human life, the life of the sciences has its periods. They are at first carried along in a toilsome infancy; now they sparkle with the days of youth; who would want to deny them those of virility? For a long time anatomy was descriptive and particular; nothing will stop it in its tendency to become general and philosophical." See *Rapport à l'Académie*, etc. This report was published in the *Moniteur* of October 29, 1829.

rather, let us do better: let us avoid both these reefs, thinking of what we owe to the sense of that adage, *in medias stat virtus.*

Who remembers today that in the first years of the revolution, classifiers following the Linnaean method, naturalists occupied only with species, came to the Royal Garden to place under the oldest of our cedars of Lebanon a bust of Linnaeus? They wanted much less to honor the greatest naturalist of modern times than to protest against the development of the followers of Buffon, whom they reproached for abandoning themselves too much to the seductions of imagination and poetry; unhappy efforts, of which posterity has taken no account! The fact is that the public, where the most diverse opinions converge, where all the needs of the classes are concentrated, and which therefore enjoys an instinctive view as certain as it is extensive, rejects as erroneous all those condemnations by the spirit of partisanship. Force and elevation of thought are necessarily stamped with imagination and poetry: the writings of Buffon are elaborate facts that prove this incontestably. We now know, today, when so many editions of the *Histoire naturelle* succeed one another so rapidly—monuments of a sort which repeat in their way and which sanction that judgment of his contemporaries, that cry of admiration that Buffon heard in his lifetime, and which lives on traced at the base of his statue: *majestati naturae par ingenium.*

FIRST SUMMARY*

OF DOCTRINES RELATIVE TO THE PHILOSOPHICAL RESEMBLANCE OF BEINGS BY THE EDITORS OF *Le Temps*

(ISSUE OF MARCH 5, 1830)

The solemn discussion that has just taken place at the Academy of Sciences between MM. Cuvier and Geoffroy Saint-Hilaire draws the attention of all educated men. Let us try to present an idea of the arguments on which each of the two scholars based his opinion.

For a decade, naturalists have been much concerned with a theory proposed by M. Geoffroy Saint-Hilaire under the name of theory of analogues, which that scholar presents as being able to offer new bases for zoology. The fundamental principle on which it rests consists in admitting that all animals, whatever the diversity of their forms, are the product *of one and the same system of composition*, and, physically, *the assemblage of parts that reveal themselves uniformly*. This principle has been received with favor in France and in some foreign countries. Learned investigations have appeared to offer more or less positive confirmations of this doctrine.

*Two contemporary journals, still more especially devoted to ascertaining the progress of the sciences and of literature than to following the discussion of politics, *Le Temps* and *Le National*, did not confine themselves to reporting, in the chronological order of the sessions the facts debated within the Royal Academy of Sciences on the subject of analogues: the writers for those journals thought that if they used less technical expressions, they would bring to a more numerous public the points in these grave questions of science which are difficult to understand. It is to their thoughtful attention to the greater number of their readers, that the public is indebted for clear and luminous summaries of these materials. I do not see that this could have been done better; the questions are handled in different ways, and in effect with such superiority, that I thought I would give pleasure by reproducing these summaries, giving them here word for word. G.S.-H.

However, the theory proposed by M. Geoffroy Saint-Hilaire does not receive general assent; it has even been rejected since its origin by a naturalist whose works honor the learned world, M. Cuvier, who has never stopped protesting against its admission, but who has abstained from combating it directly until the moment when a particular circumstance finally made him decide to enter the lists.

Among the mollusks in general, the cephalopods in particular distinguish themselves as having an extraordinarily rich organization and a great number of viscera analogous to those of the higher classes. They have a brain, often eyes which, in the cephalopods, are still more complicated than in the vertebrates; sometimes ears, salivary glands, multiple stomachs, a very considerable liver, bile, a complete double circulation, provided with auricles and ventricles; in a word, powers of very vigorous impulsion, distinct senses, very complex male and female organs, from which emerge eggs containing a fetus; and the means of alimentation are arranged as in many vertebrates.

However, M. Cuvier was far from thinking that these animals, with such a complex organization, could be regarded as formed on the same plan which appears, up to a certain point, common to all the vertebrates. He even declared formally that they seemed to him to offer the example of an essentially different system of composition, and, by that remark, made in advance against the principle of the unity of composition an objection which, if it was well founded, would entirely reverse it, since who proclaim it regard it as absolute and as not being able to allow any exception.

Until recently, none of the partisans of the principle of the unity of composition had tried to show how the organization of mollusks could be reduced to that of the vertebrates.

MM. Laurencet and Meyranx were the first who tried to undertake that difficult task. They thought they had resolved the problem by considering the mollusks as vertebrates, folded forward at the level of the navel, so that the parts of the vertebral column were brought into contact.

The memoir of these young naturalists, submitted to the judgment of the Academy of Sciences, was the object of a very favorable report by M. Geoffroy Saint-Hilaire, who, in giving his approbation to the authors' point of view, remarked that it was directly contrary to the assertion formerly enunciated by M. Cuvier, and that it furnished an interesting confirmation of the great principle on which he had no doubt that zoology would henceforth prove to be based.

It was in reply to that assertion of M. Geoffroy Saint-Hilaire, that, at the session of February 22, M. Cuvier read his memoir *On the Organization of Mol-*

lusks, in which he devotes himself to the examination of the principle of the unity of organic composition. "Above all," said M. Cuvier, "we must make our terms precise; it is necessary to know what is meant by those expressions, unity of composition, unity of plan. If you take the words in their most rigorous usage, you could not say that there is unity of composition in two kinds of animals unless they are composed of the same organs. In the same way, to be able to affirm that there is unity of plan in their organization, it would be necessary to show that these same organs are disposed in the same order in one group as in the other. But it is impossible that you should understand things in this way: that you would want to maintain that all animals are composed of the same organs arranged in the same manner. No one would say that man and the polyp have in this sense *one* single composition, *one* single plan.

"Thus by *unity,* you do not understand *identity;* but, giving to that word a sense different from what one would naturally assume for it, you are using it to signify *resemblance, analogy.*

"With the terms thus defined, your point of *unity,* confined within proper limits, seems incontestably true; but then it is far from being new. On the contrary, it forms one of the bases on which zoology has rested ever since its origin, one of those on which Aristotle, its creator, placed it; and all the efforts of anatomy have never, through all the centuries, stopped being devoted to its consolidation. Thus every day we can discover in an animal a part that we did not know was there, and which allows us to grasp some further analogy between that animal and those of different genera and classes.

"There may be connections, relationships newly perceived. The labors undertaken in this direction are eminently useful; and those of M. Geoffroy Saint-Hilaire in particular are worthy of all the esteem of naturalists; it is further traits that he has added to the resemblances of various degrees that exist between the composition of different animals. But he has only extended the ancient and familiar bases of zoology and does not appear to have proved the unity or identity of that composition—nothing, in short, that can make room for the determination of a new principle.

"Thus, to sum up, if by *unity of composition* you mean *identity,* you are saying something contrary to the simplest testimony of the senses; if by that phrase you mean *resemblance, analogy,* you are asserting a proposition true within certain limits, but as old in its principle as zoology itself."

Furthermore, as to this so important and so ancient principle, M. Cuvier—and this is above all how he differs from the zoologists he is combating—M. Cuvier is far from adopting it as unique; he regards it, on the contrary, as subordinate to another much higher and much more fruitful principle: to that *of the conditions of existence, of the conformity of the parts, of their coordination for the role that an animal has to play in nature.* Such is the true philosophical principle from which flow the possibility of certain resemblances and the impossibility of certain others, the rational principle from which the analogies of plan and composition are deduced, and in which at the same time it finds limits that it would be foolish to try to disregard.

After having thus combated in a general manner the principle of the *unity of composition,* M. Cuvier shows that the application that MM. Laurencet and Meyranx wanted to make of it cannot be admitted. To prove this, he takes, on the one hand, a vertebrate, which he had folded as the hypothesis of those naturalists required (the pelvis toward the nape of the neck), and, on the other, a mollusk placed in position; then he compares the respective positions of the parts. It results from this examination that the resemblance pointed out by the authors is entirely imaginary. Perhaps it would be a little less difficult to establish some analogy of situation, by supposing the animal folded in a direction inverse to the hypothesis (the pelvis toward the anterior part of the head).* But on this very supposition, the problem would be far from resolved. M. Cuvier

* That is exactly what, according to MM. Laurencet and Meyranx, forms the specific character of the second family of mollusks, the *gastropods.* Not to depreciate in any way the merit of those anatomists, I had been careful not to say, in my report, that I had had, in 1823, an idea a little like theirs; but in these days of lively discussion, I join them to take my part in the perils of the struggle.

I had in fact placed in the columns of the celebrated physician Broussaid, *Ann.* etc., vol. 3, p. 249, an article under the title: *Système intra-vertébrale des insectes,* where the following is found: "As for me, I have never been able to consider a tortoise, enclosed in its double carapace, without thinking that the snail is enclosed in the same way in its shell, and that, however great the difference of the two organizations, these animals succeed in this by the use of the same method, by putting into play analogous organs.

"The pectoral chamber, or, to speak analogically, the shell of the tortoise, is open at both extremities; consequently, there is no obstacle to the digestive canal's having two openings, of entry and exit, each at one end. But in the mollusks with univalve shells, there the chamber has only one opening for the mouth and the anus, the two openings of entry and exit have approached one another and are arranged alongside one another; the compound structures *diazoma* and *distoma* are in this situation. The fact is that the digestive canals have been diverted, and thus finally folded over themselves, to arrive almost at their point of departure. I am not here making a mere supposition as far as the higher vertebrates are concerned. Indeed, see how in the sole, the anus opens behind the furcular bones; it is so close to them, that the abdominal viscera are forced upward and in part flung

goes much further; he believes he can state that it is impossible that this will ever be done in any fashion, and supports his assertion with the consideration of numerous and enormous differences, less on one side and more on the other, that the vertebrates and the mollusks present.

M. Geoffroy Saint-Hilaire began the defense of his doctrine at the session of Monday, March 1. He indicated with precision what the principle is that he has supported up to now. First, he has never made the distinction between these two ideas—unity of composition, unity of plan—; and all that has been inferred in the way of exaggerated consequences to which their conjunction could lead is entirely false.

Led by observation alone to recognize that all animals are formed according to one and the same system of composition, he called the principle that expresses that insight the principle of the unity of composition; and he does not see how anyone could reasonably object to that expression. But, it was said, are you speaking of absolute identities, or simply of analogies, of resemblances? "I have not," M. Geoffroy replies, "ever meant anything beyond what those last words express." "Then you have said nothing new, and, far from having placed zoology on new foundations, as you claim to do, you are only repeating a truth known since Aristotle."

Is this assertion exact? That is what M. Geoffroy proposes to examine in his first memoir. He does not deny that Aristotle had a presentiment of the principle of the unity of composition, or that it had not been glimpsed also by some superior men, by Belon, Bacon, Newton himself; it is on the idea of analogy that the whole scaffolding of the methods of natural history rests.

"Thus," continues M. Geoffroy, "if I had done nothing but perceive similar analogies, nothing but indicate new ones, following the method adopted up to now, I would have had no right to claim priority."

But that is not how it is: first, M. Geoffroy has not been limited to receiving his inspirations from Aristotle; it is from nature itself that he has drawn them. He has interrogated the facts, attaching himself with ardor and perseverance to the pursuit of truth. He has descended to the examination of the minutest details, and his conviction is the fruit of his personal studies.

But it is not only because he has pursued his ideas with an uncommon perseverance that M. Geoffroy has come to recognize analogies where others be-

back from behind; they hollow out under the skin a box to the right and left of the anal fin. Do not, however, believe in a change of connections; this metastasis is more apparent than real. Etc., etc."—G.S.-H.

fore him had seen only differences. He has owed his successes above all to a method that is his very own, on which above all he bases the right that he finds he has to present himself as founder of a new doctrine. In fact, before him, it was almost exclusively the consideration of forms and functions that had guided naturalists in the investigation of analogies.

Far from following the same path, M. Geoffroy rejects all deduction founded on the consideration of forms and functions, and proclaims the principle that all zoological research can have no other basis than anatomy. Thus, of the three kinds of considerations on which naturalists rely in the investigation of analogies, M. Geoffroy eliminates two as entirely defective. Only one, according to him, is to be regarded as having real value; but that one suffices, not only to establish the reality of analogies that had been previously recognized, but even to make some perceptible that no one had suspected until the present, in order to establish the great principle of the unity of organic composition.

In the old philosophy, it was the organs according to functions, taken in their totality, that was considered; in M. Geoffroy's theory, it is between the materials constitutive of these organs that resemblance is to be sought.

Let us take an example: the hyoid bone of man is composed of five small bones, that of the cat of nine. Are these two parts, called by a single name, analogous in the one and in the other species? To answer that question affirmatively, in the old doctrine, it is sufficient that they be devoted to the same use; but in M. Geoffroy's doctrine, it is not like that; and the hyoid of man furnishes the analogue only of five of the parts of that of the cat.

Thus four parts are lacking in the hyoid of man, and these parts, according to the doctrine of analogues, must necessarily be found somewhere. The naturalist, alerted by this, will look for them in the vicinity of the organ that is deprived of them, and, guided by another principle of the new doctrine, that of connections, he will not be long in recognizing them in the needle-like protrusions placed on the two sides of the auditory passage in man, to which naturalists who fail to recognize their origin have given the name of *styloid apophyses*. Thus these parts of entirely different forms, deprived of the function that they fulfill in the cat, are the true analogues of a part of that organ.

To sum up: 1. M. Geoffroy has arrived at the theory that he puts forward by investigations that are his own.

2. The old school, following M. Cuvier, admits the principle of analogy within certain limits; M. Geoffroy, on the other hand, recognizes no exception whatsoever to his principle of organic composition.

3. The path that M. Geoffroy follows in his zoological studies is essentially different from that which his predecessors have adopted. They seek to establish their analogies according to the consideration of forms, according to that of functions, and, finally, after that furnished by anatomy. M. Geoffroy wants all zoological research to be founded uniquely on anatomy, and, with that sole element of research properly employed, he arrives at consequences much more extended than those to which his predecessors were confined. Nothing is less justified than the reproach that has been addressed to him of not having enlarged the ancient foundations. M. Geoffroy has incontestably attempted to reverse the bases set up by his predecessors and to establish new ones. He may have been wrong; he may have been right; that is not what it is a question of examining for the present. But, for good or ill, the path that he has followed belongs to him alone.

M. Cuvier did not think he needed to reply to M. Geoffroy Saint-Hilaire's memoir; he confined himself to remarking that everything that his learned colleague had just said might be true, without any one's being able to conclude from that anything about what he himself had put forward in the last session, relative to the impossibility of reducing the organization of certain beings of the inferior classes, that of the cuttlefish in particular, to the plan that appears to be common to the vertebrates. M. Geoffroy, he added, announces that he will take up the question later; we will then be able to discuss it.

It seems to us that M. Geoffroy should have remarked on his side, that he had established in an incontestable manner all that he had proposed to prove for the moment, to wit: that the principle of the unity of organic composition, as he understands it, differs essentially from all that had been adopted up to now on the analogies that exist among organic beings, and that he had arrived at these ideas following a path that is uniquely his own.

The honorable academician announces that he will penetrate the very foundation of the question. We shall continue to keep our readers informed of the discussions to which these subsequent memoirs may give rise.

FIRST SUMMARY

OF DOCTRINES RELATIVE TO THE PHILOSOPHICAL RESEMBLANCE OF BEINGS BY THE EDITORS OF *Le National*

(ISSUE OF MARCH 22, 1830)

Questions of the greatest interest are at the moment the object of an ordered discussion within the Academy of Sciences, between two naturalists of the first rank, M. Cuvier and M. Geoffroy Saint-Hilaire. It is a question of nothing less than of whether zoological philosophy, as Aristotle created it, as the work of twenty-two centuries has carried it on, and, finally, as M. Cuvier himself has crowned it with admirable work that has placed him without contest at the head of the naturalists of our era—whether that philosophy, we say, shown to be insufficient and incomplete, will give way to doctrines recently introduced in zoology and comparative anatomy in Germany and France by several learned scholars, among whom M. Geoffroy occupies a very high place. When scientific discussions concern only works of detail, they remain enclosed in the womb of the academies and learned societies. But when they bear on the highest generalities of a whole science, when there can result from their shock one of those revolutions that count in the history of the human mind, when they are engaged in and sustained by men with a European reputation, then public curiosity is awakened and attaches itself to them. By repercussion, all the sciences are challenged, and have a major interest in the result. The controversy raised between M. Cuvier and M. Geoffroy Saint-Hilaire offers all these characteristics. The public could not remain indifferent to it. The questions at issue are such that, independently of their purely scientific interest, they are, beyond that, of a nature to seize the imagination of every thinking man, and to lay a strong hold on all intelligences for which the spectacle of animate nature is a fruitful source of emotions, poetical, philosophical, or religious. But there is no soul so little cultivated and well organized that it has no such feelings.

In writing on this subject, we make no claim to substitute ourselves for our scholars in the exposition of their ideas. Both of them, each according to his gifts, speaks a language that they both understand, and that the public also understands. All we want, through some preliminary and less technical explanations, is to allow them to be heard and understood by a more numerous public.

We shall try to give as clear an idea as possible to anyone who has made no special study of the anatomico-philosophical doctrine of M. Geoffroy, known under the name of *theory of analogues*. Without previous knowledge it would not be possible to follow easily the discussion that was opened on this subject, a propos of the first memoir of M. Cuvier, read at the session of February 22, which contains a criticism of it. In fact, these two naturalists, addressing a public fully informed of what was at issue, neglected with good reason many of the antecedents and explanations necessary for most of their readers.

M. Geoffroy's system, very extensive, very complex, is deduced from an infinity of the most difficult anatomical observations, which it is impossible to recall and even to refer to in this brief analysis. Thus we shall present only the most general results, which all the world can grasp, since, like all theories, this one is definitively reducible to three or four very simple propositions.

The number of animals distributed on our globe is immense, whether they live in the air or in the water, in the interior of the earth or at its surface. The number is still undetermined for us, since every new instrument, added to our organs, discovers new ones. A powerful microscope allows us to see thousands in a few ounces of liquid. The simplest attention shows that these innumerable beings resemble one another in certain respects and differ in others. All the tongues of all the peoples confirm this observation. The first classifications were probably made by fishermen and hunters: they are still employed in ordinary language, and always will be. They bear on the most evident and most striking of analogies and diversities of organization, and suffice for the needs of life and of practice. But science is more demanding. It wants more rigor in its classifications, and rules that do not allow exceptions. Comparative anatomy has discovered in the structure of animals a multitude of relations and varieties. From these accumulated observations were born zoological *methods*, which consist in classing animals in several groups, designated by the names of *genera*, *orders*, *classes*, *species*, *varieties*, etc., and in distinguishing among them by the physical characteristics that some possess to the exclusion of others.

The simplest, like the most learned, classifications, are pure abstractions by the mind, which, neglecting the differences, consider only the points of analogy. Nature, as has been said with profundity, creates only individuals; it is we who create species, by abstraction from the diversities and by combining the resemblances, a combination on which we impose a collective name. The difficulty consists in marking the analogies and the varieties well, and this difficulty is great enough to make the naturalists who are concerned with this arrive at different results: the classifications are also very numerous and often based on opposing principles. However, there are some which, although very ancient, keep reappearing in science, and are still flourishing today. Such is the classification of Aristotle, consolidated by Linnaeus, and adopted in our day by MM. Cuvier and Lamarck, although under other names.

The great concern of naturalists at all times has been to arrive at a perfect classification; that is to say, at a classification based on the complete knowledge of the resemblances and the differences of all the beings of the animal scale, and to determine these relationships with precision and clarity.

Comparative anatomy, which alone can furnish the elements of this problem, took a new direction toward the beginning of this century. Naturalists had always thought, and a great number today still think, that animal species are each provided by nature with particular, special organs, in conformity to the ultimate role that they are destined to fulfill. They see well enough that all the beings on this great scale offer some general resemblances; but the differences between certain classes are so enormous, so decisive, in their opinion, that it is impossible to admit that they were all created on the same plan. Thus, for example, the bird, which breathes in the air and flies, has different organs and different apparatuses from the fish, which breathes in the water and swims. The life of these beings is so different that a different organization has also been needed to make this possible. If we descend to the invertebrates, and if we compare them with vertebrates, every appearance of analogy disappears. These are new beings, constructed on a special model, composed of particular organs, which they, to the exclusion of all others, possess.

This doctrine has been generally adopted by philosophical naturalists from Aristotle to our day.

Since about thirty years ago, other principles have been introduced: in Germany, through the work of Kielmeyer, Oken, Spix, Tiedemann, F. Meckel, etc., and also through the speculations of the school of *nature*, in France, through the writings of M. Geoffroy Saint-Hilaire, and through our communications with Germany.

These new ideas of anatomical philosophy are not entirely the same in

France as in Germany; but one can see that they have significant enough relationships, and issue in almost the same theoretical results. It is by analyzing M. Geoffroy's own doctrine that we will indicate the spirit, the aim, and the principles of this philosophy; for, in France, M. Geoffroy is its most powerful protagonist, and he has imprinted on it an originality and a remarkable character.

M. Geoffroy's doctrine is particularly well known, and is designated by him, under the name of *theory of analogues*. In fact it consists entirely in the idea that he has formed for himself of the relations of *analogy* established between all the beings of the animal creation. It is also in defining clearly what he understands by the word *analogy*, and by explaining the means by which he identifies it, that we shall have an adequate idea of his whole system.

According to M. Geoffroy, classificatory naturalists have been much more interested in differences than in analogies in their comparative studies; and the reason is that they have not compared the organs of animals except in the context of *their form and their uses;* they saw the analogy only when it was manifestly characterized by resemblances of structure and by the functions of the parts. As soon as this resemblance failed them and was effaced, which happens soon after one passes from one species to another, they thought themselves in the presence of *new* objects, and, consequently, they also imposed new names on them. This difference in names made them see everywhere a difference in the things, and the analogy was lost from view. Thus the veterinarian, looking at the forelimb of an ox and noticing that its *form* differs considerably from that of the human arm also designates differently all the parts that compose it. He calls canon bones, spurs, hooves, the parts that in man carry the names metacarpal, rudimentary fingers, nails. The lower extremity of the forelimb of this ox, or in other words, the foot, compared to the extremity of the same limb in the ape, is no longer a *foot,* if we pay attention only to the form and to the use, but an entirely *different* organ, which is also called by the different name, *hand.* In the lion, this foot is a claw, in the bat a wing, in the whale a flipper: so that in giving a different name to the same organ, and attaching a different idea to each difference of name, the principle of analogy is obscured and finishes by being totally neglected.

Thus it is by no means on considerations of forms and of functions that zoology could have found analogies between species and reduced animal organization to a common type. If this analogy exists, it exists elsewhere. Forms and uses of parts not only change in every species, but even in every individual; indeed, it is on these two circumstances of organization that all the apparent varieties of animals depend; they are the very principle of variety. The principle of analogy or of unity is elsewhere. M. Geoffroy has called it the principle of *connections,* and this is what it consists of:

Every organized body is composed of distinct parts, arranged in a certain order in relation to one another.

Anatomically, we have to consider in every animal, on the one hand, only the form and volume of the parts and, on the other, their number and their reciprocal arrangement. Since the principle of unity and analogy that we are seeking is not found, except to a certain degree, in the form, it can be encountered in a complete fashion, if it exists, only in the order established between the parts. It is, in fact, in this order that M. Geoffroy has found it, clothed, according to him, in the highest character of generality and authenticity. Thus it is by no means the organs that resemble one another, but the *materials* that compose them. These materials themselves resemble one another neither by their form nor by their use, but by their number, their situation, their dependence on one another; in a word, by their *connections. The law of connections* admits neither caprice nor exceptions; it is invariable. We find in each family, each species, all the organic materials that are found in the others. The body of the ape, of man, of the elephant, of the bird, or of the fish is composed of a certain number of pieces placed, in relation to one another, in the same arrangement. Thus the forelimb of the horse, compared to the upper limb of man, offers only a gross analogy according to consideration of the form; but there are, from one side and the other, the same bones, the same articulations, the same muscles, the same disposition and relations among all the parts; that is to say, the same connections. To form animals, nature has only a limited number of organic elements, which it can shorten, diminish, efface, but not transpose from their respective places. It is like a city, for example, the plan of which, made by avarice, has traced the streets and the number of the houses; the architect can, indeed, alter to infinity the form of the dwellings, their dimensions and their intended uses, but he cannot disturb the order prescribed in their arrangement. That order, that arrangement, those *connections,* are always identical in all animals. Strictly speaking, there are not a number of animals, but one single animal, whose organs vary in form, use, and volume, but whose constitutive materials always remain the same in the midst of those surprising metamorphoses.

And these metamorphoses themselves, from which the differences arise, are explained by another principle, another law, which M. Geoffroy has called the *balancement of organs.* It is a law by virtue of which an organ never assumes an extraordinary development without another organ's undergoing a proportional decrease. In the regular and normal state, it is this unequal distribution of material that causes the astonishing variety of animal forms. The theory of *monstrosities* is founded on this law and obeys it. *Monsters,* which were so long regarded as strange caprices of nature, are only beings whose regular development has been arrested in certain parts; and—a thing to be wondered at—it never happens to an organ that it loses, in an individual case, the normal characteristics of the species to which it belongs, without that deformation's imprinting on that organ the normal characteristics of a lower species. It is the same for the natural development of animal bodies. Thus, man, considered in his embryonic state, in the womb of his mother, passes successively through all the degrees of evolution of the lower animal species: his organization, in its successive phases, approaches the organization of the worm, the fish, the bird. It presents temporarily all the organic combinations of which nature is so prodigal; but it does not conserve them: it sheds them, to pass on to others, until finally it arrives at that which is specially and irrevocably assigned to it. What is true of the whole animal body is again true of each of its organs. The human brain, for example, undergoes a sufficiently great number of changes, each of which has its permanent model in the brain of reptiles, fishes, etc. Tiedemann, in Germany, and M. Serres, in France, have specially noted these laws of formation.

As we said, there are not a number of animals, but one single animal, whose constitutive pieces are necessarily the same in all species, despite the numerous varieties of form that their unequal development imprints on their components. These components themselves, that is to say, the organs, do not change their nature when they change their names. Take, for example, the *sternum,* a bone situated in man at the front of the chest, and whose function is to serve the movements of respiration and to protect the delicate organs that it covers. If we compare this bone, solely in relation to its general form, to the part that takes its place in other animals, we lose the thread of the analogy, and see different organs. M. Geoffroy, basing himself on its situation in relation to adjacent organs, understands by *sternum* an assembly of pieces that form the lower part of the chest, and which enter necessarily into its composition, whether by assisting the mechanism, or by protecting the respiratory organ against external attacks. Thus the word *sternum* is a collective word, designating an assemblage of diverse bony parts, each of which, following its respective degree of development, contributes in a special way to the general uses of the whole organ, which they constitute by their conjunction. Thus we are led to an ideal type of *sternum,* which, for all vertebrates, is resolved into various secondary forms, following the variation of its material constituents. It is the same for the foot, the hand, the skull, etc.: there are not as many skulls, feet, hands, as there are animals. Just as there is only one animal, there is only one sternum, one foot, etc. In short, whatever are the particular metamorphoses of these organs, it is not difficult to disentangle their diversities, to perceive that they are converted into one another, to include all their common points, and to reduce them to one and the same measure, to identical functions, in a word, to one and the same type.

Each organic system that has attained, in a given species, its maximum of development, and hence of function, retains, with a fixity of number, the rank and the uses of its elementary portions, while in another species, in which it exists only in the embryonic state, it is exposed to the loss of its importance and its uses, and even to the diversion of some of its pieces to the benefit of neighboring organs. But whatever means nature employs to operate enlargements at one point and decreases in size at another, no one part, by a law that is imposed on it, ever encroaches on another: an organ is rather diminished, effaced, annihilated, than transposed.

By the *connections,* we arrive at the law of unity of identity of organic forms. By the *balancement of organs,* we explain their varieties and their apparent differences.

Thus *the principle of connections* and that of *the balancement of organs,* each explained by the other, lead M. Geoffroy to this conclusion: that animals are created *on the same plan;* that there is, for the animal kingdom, *unity of organic composition;* and this conclusion is the corollary of the more general one of the *theory of analogues.*

Such is the philosophical doctrine of M. Geoffroy Saint-Hilaire.* It seems,

*A reproach directed with great insistence by the preceding arguments against the author of this doctrine is a kind of pretension to the universality of his views. However, once the investigations were undertaken, what other conduct would they have prescribed for him? One is not accepted in the sciences for announcing an abstract proposition, for which it is then necessary to enumerate the exceptions. *There is no rule without exception* is a common enough locution; but it is no less an inadmissible antilogism: for the exception destroys the rule, or sometimes confirms it only when the obstacle that falsifies it appears manifest.

The universality of the principle of unity of organization is a necessary fact, and this necessity already merits demonstration. And in fact, when all the arrangements of the universe are considered on principle, it turns out that to a very small of materials, there are applied, for their disposition, forces that are, numerically speaking, also restricted: forces that are themselves only the reciprocal and at the same time simultaneous action of the properties of elementary bodies.

as he himself says, to be the confirmation of the principle of Leibniz, who defined the universe as *unity in variety.*

M. Geoffroy has not yet applied the method of determination of the organs by connections to all the classes of animals, but only to the four classes of *vertebrates,* and to the *articulates.*

Questions of priority have often been raised concerning M. Geoffroy's ideas. Some have claimed that, though new among us, they were already old in Germany. Others, and in particular M. Cuvier, maintain that they are not new, neither in France nor in Germany, but that they date back two thousand years, and have nothing new but the name. Questions of priority are always difficult to resolve. What is certain is that in 1796, that is to say, thirty-four years ago,* M. Geoffroy expressed clearly, in our opinion, the fundamental principles that he still upholds today; and, searching in Germany, we find at that date no well-known work that contains them. Thus nothing prevents us from regarding

The creative power, by combinations that are also simple, produced the actual order of the universe, when it attributed to each thing its proper quality and degree of action, and when it had ruled that such and such elements, thus emerging from its hands, would be eternally abandoned to the play, or, better, to all the consequences of their reciprocal attractions. G.S.-H.

* *It is to the following passage that this reflection is alluding:*

"A constant truth for the man who has observed a great number of the productions of the globe, is that there exist among its parts a great harmony and *necessary* relationships; that is, that it seems that nature has restricted herself within certain limits, and has formed all living creatures only on one unique plan, essentially the same in its principles, but which she has varied in a thousand manners in all their accessory parts.

"If we consider one class of animals in particular, it is there above all that its plan will appear evident to us; we will find that the diverse forms under which nature has been pleased to make each species exist, *all derive from one another: it is sufficient to change some of the proportions of the organs, to make them fit for new functions, and to extend or restrict their uses.*

"The pocket of the howler monkey, which gives that simian its resounding voice, and which is perceptible at the front of its neck by such an extraordinary hump, is only a swelling of the base of the hyoid; the pouch of the [didelphes], a folding over of their skin, which has great depth; the elephant's trunk, an excessive prolongation of its nostrils; the horn of the rhinoceros, a considerable mass of hairs which cohere together, etc., etc.

"Thus the forms, in each class, however varied they are, all result at bottom from organs common to all: Nature has refused to employ new ones. Thus all the differences, even the most essential, which distinguish each family in a given class, come only from another arrangement, another complication, in short, a modification of the same organs." See "Dissertation on the Makis," *Magaz. encycl.,* vol. 7, p. 20.

M. Geoffroy as their author, at least in our country, and, if they have some philosophical grandeur, from giving that honor to France. The question of novelty need not occupy us further; for ordinarily this is an objection that is made only when many others have already been exhausted. Further, we believe that a principle injected into a science would never produce a great movement if it differed only in name from the received principles. Finally, we shall add that any principle whatsoever can find itself tucked away in twenty passages of old books, without our having to regard it as old. In short, a principle is nothing if it is not worked with and applied: it is a glimmer, a flash, a presentiment, as they say; but it has a value and a character only in the hands of the man who makes it recognized for what it is, and who proves why it is. Moreover, he alone can be regarded as its proprietor, since he alone knows what he has, and recognizes that he has it.

We are far from having exhausted this interesting subject, and we should have liked to give a greater degree of clarity to this brief exposition. We shall return to it, perhaps in another article, where we will investigate in what and up to what point the opinions of M. Geoffroy Saint-Hilaire and M. Cuvier differ.

6

OCTOBER 1830: THE CONTROVERSY REAPPEARS

For many historians, the controversy is finished on June 7, 1830, the session during which Geoffroy Saint-Hilaire presents to the Academy his work which, announced in the prospectus distributed on April 5, is printed on April 15. However, after the forced truce due to circumstances, he wishes to resume the hostilities.

Let us return for a moment to the course of history: the July Revolution, known as the Three Glorious Days, the abdication of Charles X, the accession of Louis-Philippe.[1] Cuvier, who, in the wake of 1789, "*fears the populace,*" takes refuge in London as soon as the trouble begins. Geoffroy Saint-Hilaire remains in Paris, and here he is, he, the agnostic, once more in the role of protector of the clergy, sheltering the archbishop of Paris in the Botanic Garden. Then, toward the end of the summer, everything returns to order; Cuvier comes back from London, and the Monday sessions are resumed at the Academy.

The session of October 4, 1830, interests us in the highest degree. On that occasion Geoffroy Saint-Hilaire reads a memoir entitled *Sur les lames osseuses du palais dans les principales familles d'animaux vertébrés, et en particulier sur la spécialité de leur forme chez les crocodiles et les reptiles téléosauriens* (*On the bony plates of the palate in the principal families of vertebrates, and in particular on their special form in the crocodiles and the teleosaurian reptiles*).[2] No one could be deceived. Here again are the crocodiles and *Teleosaurus*! Thus Geoffroy Saint-Hilaire is taking up again the discussion of 1825, by bringing the conflict into the field of paleontology; now, it is explicitly a question of evolution.

In his introduction, we find again one of his favorite themes. It is necessary to study fossil reptiles in depth, for their very presence poses a question: "*If the alleged crocodiles of Caen and of Honfleur contained in similar formations those of the Jurassic, with the Plesiosauruses, were they not in the order of time, as well as by the degrees of their organic composition, a connecting link that would join without interruption these very ancient inhabitants of the earth to the reptiles currently living and known under the name of gavials?"* At the same time he announces here his project of writing a series of memoirs in order to study this difficult problem. As a matter of fact, four other communications to the Academy will follow on October 11, 1830, and March 28, May 9, and August 29, 1831. (See note 2 above.) The academic year 1830–1831 is thus entirely devoted to crocodiles!

But why does Geoffroy Saint-Hilaire return to the "affair" of the crocodiles of Caen? In fact, he is here reopening a favored subject; he is returning to an area that he knows well, that of the crocodiles; he can once more demonstrate that Cuvier is not infallible, even in his special field; he can attack him on one of his basic ideas, the fixity of species; finally, he is trying to connect, with yet more precision, the transformation of species with teratogenesis. Thus we find mingled, in these memoirs, a real animosity against Cuvier and a no less real scientific curiosity.

We may ask ourselves whether Geoffroy Saint-Hilaire is abandoning the ground of the recent controversy because he thinks his position is indefensible, or whether he believes that the arguments of that old debate are directly related to those of the discussion of February and March. The chronology of these works and their natural tendency to unification clearly argue in favor of the second hypothesis. In fact, given the continuity of Geoffroy Saint-Hilaire's thought, Cuvier's ferocious opposition to experimental embryology, and the relations between embryology and classification, as they would henceforth be evident to our protagonist, to treat of paleontology was only another way of broaching what is still the same subject, that of the unity of organic composition.

Thus there is no further need to ask whether the controversy between Cuvier and Geoffroy Saint-Hilaire had or did not have a bearing on evolution.

Those who focus solely on the texts of the quarrel of February to March can answer: *"Not explicitly."* But those who have followed the tensions between the two protagonists—from 1798 to 1832—have reason to say that evolution progressively enters the basic texture of the conflict, and that the year 1830 is only one incident in a vast debate.

The second memoir, that of October 11, 1830—"The special forms of the hind-skull in the crocodiles, and the identity of the same organic parts in the teleosaurian reptiles"—is without doubt the most important. After showing the original characteristics of *Teleosaurus,* which permit its distinction from extant crocodiles, Geoffroy Saint-Hilaire studies its hind-skull, which, for him, is totally crocodilian, and which gives the *"crocodilian stamp par excellence."* He specifies in detail the anatomy of the auditory bulla, not without again raising with delight the inaccuracies of Cuvier's description. He is clearly using the law of connections to show the originality of the position of the auditory bullas, which come to be fused behind the occipital, insisting on the fact that, without this law, such a bony structure would be difficult to understand. He then at once compares this structure with a case of human teratology—the sphenencephalon—which also presents an "*auricular band*," but passes under the brain, and not above it, as in the crocodiles. Then, asking himself why we have a monster in one case and not in the other, he ascertains that it is "*uniquely this circumstance of incapacity for the life of relations*" that gives rise to that appellation. Once again Geoffroy Saint-Hilaire finds a constructive reciprocity between zoology and embryology, which will lead him to propose a theory of evolutionary processes at the scale of the animal kingdom.

Certain elements of the scientific text have something surprising about them if we do not have present in our minds the principle of correlations dear to Cuvier. In fact, Geoffroy Saint-Hilaire, returning to his earlier writings, makes explicit the double originality he sees in the crocodile—the particular structure of the auditory bullas, but also the astonishing size of the cranio-respiratory canals—and then argues at length around the fact that "*these two great modifications, which have been able to agree with one another, and which, adjusting themselves to one another in the crocodiles, become a double item indicating their isolation as a family, are nevertheless not in the least in reciprocal dependence on one another.*" Thus: *"The independence of these two abnormal combinations exists in fact; it is revealed to us by the organization of the fossil saurians of the limestone of Caen."* For anyone who can read between the lines, here is a flagrant case in which Cuvier's principle of correlations cannot apply.

But how have such organic modifications been able to occur? Geoffroy Saint-Hilaire then postulates an important action of the environment, recognizing at the same time that for the moment the inner mechanism is beyond our knowledge. For him, the environment—and, in particular, the atmosphere—of remote epochs could not be identical to the one we know today. As he postulates in his fourth memoir, it is chiefly variations in the amount of oxygen that would have oriented the evolutionary processes.[3] Geoffroy Saint-Hilaire then imagines a history of life on earth, given its rhythm by the modifications of the atmosphere, and gradually leading organisms to their present forms. We can immediately make the connection with his embryological experiments of 1826.[4] In this passage the catastrophic thesis of Cuvier is refuted, but equally the mechanism of *habit* postulated by Lamarck.[5] Finally, in his fifth memoir,

Geoffroy Saint-Hilaire defends once again the position of those who do not stop at the mere facts, but try to integrate them in a theory. Moreover, for that purpose he has recourse to a parable that points directly to Cuvier.[6]

This time Cuvier's replies no longer take place at the Academy, where he now feels himself to be less powerful than before,[7] but at the Collège de France, which produces a kind of vaudeville cross-talk act.[8] Ampère came to hear Cuvier, but went to see Geoffroy Saint-Hilaire, in order to prepare a reply, which he gave in his own course at the Collège de France, where he gave alternate lectures in physics and biology. Frédéric Cuvier came in his turn to spy on Ampère and reported to his brother, who replied the following week. This game stopped brutally with the death of Georges Cuvier on May 13, 1832.

Still, can we say that the controversy stopped at that moment? We see Geoffroy Saint-Hilaire continuing his work in evolutionary paleontology, and trying—as a matter of fact, very awkwardly—to impose his ideas and to take Cuvier's place in the scientific community. But then he finds himself facing Frédéric Cuvier, who is at the head of the pack of all the former students and collaborators of his brother Georges.[9]

Nevertheless, we should comment on some texts which correspond, from 1833 to 1838, to a certain culmination of the thought of Geoffroy Saint-Hilaire, corresponding to his desire to have a social influence.

Most of his conceptions in evolutionary paleontology are summarized and given precise form in an article of 1833, the object of which is to take account of the study of numerous fossils that a certain Abbé Croizet, an amateur paleontologist and the curate of Neschers, a village situated between Clermont and Le Puy-en-Velay, had presented to him after a voyage in Auvergne.[10] This collection had great importance, insofar as it permitted Geoffroy Saint-Hilaire to affirm his thought in this domain. Two ideas stand out. First, he reaffirms the nonfixity of species, which can vary, according to him, under the influence of changes in the ambient medium: "*The species is not fixed and does not reappear in its forms, similar to its parents, except for the reason that the conditional state of its ambient medium is maintained; for, according to the bearing and under the influence of the variations of the latter, there are scarcely any changes that are not possible with respect to it.*" He cites once more, but with more precision, his work in experimental embryology, postulating that, if there are two zoologies—that of fossils and that of now extant species—this would be because of an important change in environmental conditions. Such a change would have been produced gradually, and the passage from one zoology to the other would have been obtained without a break.[11]

The second idea concerns the origin of man. Geoffroy Saint-Hilaire considers it the ineluctable result of a concatenation of causes produced in a deterministic manner.[12] More precisely, relying on his work in experimental embryology, he postulates that the variations of organisms are directly dependent on the changes of the environment; thus it would be the latter that would determine the conditions—and thus the moment—when a particular animal, in this case man, would have the possibility of manifesting itself. The more complex the organism, the more the conditions of life have to be elaborated, hence the later the opportunity of existing will arise for it. Thus, for Geoffroy Saint-Hilaire, it is not really organisms that have a history: it is the earth that has one, producing according to its development the conditions that are going to permit embryos to actualize step by step the potentialities that already had an immanent existence.[13]

This conception of things amounts, in the mind of Geoffroy Saint-Hilaire, to the culmination of his scientific thought. In fact he is here utilizing, in synergy, the whole of his work: unity of organic composition, law of connections, principles of teratology, results of experimental embryology . . . Serres is not mistaken in this, and it is this synthesis that he will summarize faithfully at the time of the eulogy that he will pronounce at the obsequies of Geoffroy Saint-Hilaire (see below). One can also understand why Serres will later refuse to accept the Darwinian conception, not being able to accept the fact of indeterminism and contingency.

Such a globalizing vision carries with it as a corollary the notion of progress. The temptation is then great not to limit it to animals alone. And why not consider, by analogy, a progress of humanity? We find the premises for this in a note that refers to Pierre Leroux (1797–1871). The latter, relying on the evolutionary paleontology of Geoffroy Saint-Hilaire, postulates "*the indefinite perfectibility of humanity.*"[14]

It is with the same logic that, after his memoir of 1833, Geoffroy Saint-Hilaire publishes an introduction to a book by Philippe Buchez (1796–1865) entitled *Introduction à la science de l'histoire, ou Science des développements de l'humanité (Introduction to the science of history, or Science of the developments of humanity)*.[15] The tenor of the work is clearly in the spirit of the time. It is a question of taking the important notions of the sciences of nature, and applying them by analogy to the history of humanity.[16] "*Geology, embryology, and comparative anatomy are the sciences on which the author, trained and prepared by the study of their specialties, draws, by explaining one by the other, associating them in his mind, and making of them a common knowledge, to arrive at the most general facts from which he composes his humanitarian physiology . . . Humanity, says the author, would be a function of the universe, and consequently a function of a univer-*

sal law. Through observation of the past, one could succeed in knowing the law of humanity . . . The idea of progress is the Word, is the Reason of the Universe." Thus, just as the animal kingdom has passed in a perfectly deterministic manner through a succession of states generating different zoologies, in the same way humanity has passed and will pass through different phases, which—they, too—are determined. To know the law of the sciences of nature necessarily gives access to the law of humanity, which finds itself to be a function of the universe. We can imagine the sarcastic remarks that Cuvier would have offered!

In fact, Geoffroy Saint-Hilaire finds himself with a double goal: to determine that universal law, and to make contact in progressive circles in order to establish his social influence. He makes a frontal attack. Thus it is that the last publications of Geoffroy Saint-Hilaire, from the years 1835 and 1838, are marked by the development of the law *"of attraction of itself for itself,"* glimpsed as early as the Egyptian campaign, first applied in embryology for the explanation of Siamese twins, then generalized by analogy in a grandiose vision for the whole of the Universe.

The search for social action leads him to frequent progressivists like Victor Considérant (1808–1893), who, under his influence, breathed to Charles Fourier (1772–1837) the concept of *"social evolution."* He will become the intellectual master of young Saint-Simonists, certain of whom are friends of his son Isidore, like Jean Reynaud, Pierre Leroux, and Charles Didier.[17] Edgar Quinet, the translator of Johann von Herder (1744–1803), would become a friend of the family. He would have several influential proselytes, such as Frédéric Gérard (1806–1857), for who would be an editor of Charles d'Orbigny's universal dictionary of natural history, and who would assist greatly in popularizing the ideas of Geoffroy Saint-Hilaire among an enlightened public.[18]

The influence of the biologist became more and more important among numerous intellectuals, in particular among certain major writers like Honoré de Balzac (1799–1850) or George Sand (1804–1876), whose works still continue to make known the importance of Geoffroy Saint-Hilaire. Thus it is that in the preface to the first part of the *Illusions perdues (Lost illusions)*, Balzac—who, at that moment, is close to Buchez—explains that he has conceived the study of humanity as a zoology: *"When a writer has undertaken a complete description of society, seen in all its faces, seized in all its phases, starting from this principle: that the social condition so adapts men to their needs and deforms them so well that it has created as many* species *as there are professions—that, in short, social Humanity presents as many varieties as Zoology—should one not give credit to so courageous an author for a bit of attention and a bit of patience?"*[19] Here Balzac takes as given the influence of the environment on the embryology of organisms, and, by a bold transposition, postulates that, for human psychology, the social environment has a comparable influence.

In his celebrated *Introduction à la comédie humaine,* which appeared five years later, Balzac refers explicitly to Geoffroy Saint-Hilaire, when, on the very first page, he announces the comparison he wants to make between "*Humanity* and *Animality.*"[20]

As for George Sand, the relations that Geoffroy Saint-Hilaire tried to establish with the priestess of Nohant were very soon enmeshed in conflict. It seems that George Sand was working for some time on a project for a novel intended to explain the ideas of Geoffroy Saint-Hilaire. However, the materialism of the one came to shock the mysticism of the other.[21] They could not understand one another. Yet this attempt bears witness to the impact that ideas stemming from embryology and anatomy could have on the enlightened minds of the time.[22]

This last adventure was Geoffroy Saint-Hilaire's swan song. His health deteriorated increasingly. In July 1840, he lost his sight, probably a consequence of his voyage to Egypt. In 1841 he left his teaching posts. He would have passed forty-seven years as professor at the Museum. He was now an old man who was no longer active scientifically. Despite all this, he kept up with the principal discoveries in biology, chiefly through the intermediary of his son. He died on June 19, 1844.

Geoffroy Saint-Hilaire's funeral services brought together the whole of the intelligentsia and were the occasion of various speeches which allow us to measure his importance. *La gazette médicale de Paris* reported the obsequies in full. As they inform us:

> The pallbearers were MM. Baron Charles Dupin, president of the Academy of Sciences; Duman, dean of the Faculty of Sciences; Ferrus, president of the Academy of Medicine; and Chevreul, director of the Museum of Natural History. The cortege was numerous and composed of scientific, artistic, and literary notables. Among those noticed were MM. Arago, Flourens, Duméril, Mathieu, Jomard, Dupin, Elie de Beaumont, Dufresnoy, Rayer, Cauchy, de Blainville, Poncelet, Ballanche, Pariset, David, Victor Hugo, Renaud, Guinet, etc.; a deputation from the Normal School; and a crowd of physicians, of savants, and of people from all classes of society.

Appended here are three of the funeral orations, those of Dumas, Serres, and Quinet.

FUNERAL ORATIONS

ORATION OF M. DUMAS, DEAN OF THE FACULTY OF SCIENCES

Gentlemen:

Twenty years ago, when a stranger entered the sanctuary of the Academy of Sciences, he remained there with his mind struck by a sacred respect when he saw united around him so many rare geniuses, the pride of France and the luminaries of their century.

The geometricians bowed before Laplace, Ampère, Legendre, Poisson; the physicists came to honor Haüy, Berthollet, Chaptal, Vauquelin; and the naturalists proudly opposed to those illustrious and popular names those of Jussieu and Desfontaines, those of Cuvier and Geoffroy Saint-Hilaire, also long consecrated by the veneration of Europe.

All these men, whom the republic had elevated, whom the empire had rekindled and enlarged to the height of their glory, all these men, of iron for work, and of fire for thought, whose discoveries have supported without fading the luster of the greatest events of war and politics, all these men, alas, have disappeared from our midst, and this open tomb will soon close on one of the most illustrious among them. In this mourning of science and of our fatherland which the Academy has renewed, with so many repetitions, in less than twenty years, may our souls rise to the height of the geniuses that are extinguished! May the youth who hear me understand that, in the absences of those great social crises that support in man's hearts a sacred dismay and a salutary ardor, we can find in duties that are calmer, but fulfilled with conviction and faith, equally happy inspiration.

It is to the Faculty of Sciences that it belongs to announce this. The illustrious author of the *Anatomical Philosophy* found in the demands of his instruction the occasion and the source for the most curious developments of his system. Indeed, attached to the cabinet of natural history in 1793, Geoffroy Saint-Hilaire counted more than a half century of service in that fair establishment; but, assigned to a course that included only the mammals and the birds, France would count fewer among the great discoveries that constitute her glory, if the creation of the Faculty of Sciences had not opened for Geoffroy Saint-Hilaire a new theater for a more general and more advanced study.

From that moment, his thought, supported by the respectful attention of distinguished students, distinguished above all by their philosophical studies, leapt more freely into the field of abstraction, and succeeding in determining the laws of animal organization to which his name will always be attached, and of which he had already been long cognizant.

Thus there are two periods in the life of Geoffroy Saint-Hilaire. In the first, which runs from 1793 to 1807, he shows himself to be an incisive zoologist, gifted with a large and sure eye.

The anatomist's future discoveries do indeed exist in germ in the memoirs of zoology that he published during those fourteen years; but these germs remain unknown, and it is to the zoologist that the Academy of Sciences opened its ranks in 1807, and on the zoologist that the University conferred the chair of the Faculty of Sciences in 1809.

Until then, anatomical philosophy as he has conceived it did not exist. It is with us, for us, I would even say through us, that he has established that discipline. As he tried every year to surmount new difficulties, fortifying himself in his convictions by new proofs, confirming himself in his doctrines by their success even with the young, it came about that his doctrines were established, little by little, on foundations that could be thought unshakeable.

Thus, when our venerable colleague, smitten by age and by illness, and already blind, came a few years ago to put his resignation into my hands, I rejected it with all my power; in his strictness, he wished, he said, to leave free a chair he could no longer occupy; and as for me, I begged him to preserve a name that had contributed so much to establishing the glory of the Faculty of Sciences, and I was glad of my success in convincing him. It was illness alone that could have snatched him from duties that he fulfilled with enthusiasm: sure of having to spread new ideas and great ideas, and always ready to make himself at once their apostle and their martyr.

For Geoffroy Saint-Hilaire had long understood well that he must accept the role of head of a school, and that he must submit to all those duties, and all the bitterness of that position, so much envied by those who are ignorant of its secret sorrows, and who see only the external luster.

It is with this salutary preparation of public instruction that he faces the discussion on his principles raised in 1829 by G. Cuvier within the Academy of Sciences.

Relying on numerous observations, cleverly grouped, from which flowed clear, precise, pressing deductions, Cuvier, confident of his genius, so often demonstrated, could count on an assured triumph.

Geoffroy Saint-Hilaire's less numerous observations perhaps lacked that precision desired by a hand skilled in the art of dissection; they were less well arranged and described, of a colored style, but lacking the sharpness that the sciences demand, at least in our country. Thus in terms of form everything was against Geoffroy; nevertheless the public, with its admirable instinct for the truth, was not deceived.

From the first day of the debate, everyone took to hoping that the views of Geoffroy Saint-Hilaire would be confirmed. Everyone understood that the human mind was going to take a great step.

Cleverer and more eloquent voices will tell you the causes of this favor; they will show Geoffroy Saint-Hilaire rising again to the general principle of the unity of composition, of the unity of type, of being belonging to a given organic system.

They will tell you how these beings, so similar in the origin of their development, later diverge through the cessation, the aborting, or the excessive development of certain organs, nature thus creating the apparent difference beneath which a skilled hand discovers hidden analogies.

Thus, alongside the investigation of facts, of their exact and attentive observation, Geoffroy Saint-Hilaire placed the investigation of the causes that obtain between them, which permit the establishment of interconnection and necessity, and, to use here the expression of a judge whom I have no right to praise, M. Villemain, science thus understood constitutes the first of philosophies. This unity of type, which serves him as the base for tracing all the facts of comparative anatomy, has now been laid hold of by the science of plants and has been able to encompass it with the most convincing demonstrations. Now it is penetrating the chemical sciences and is perhaps preparing a revolution in ideas in that field.

These views, so great, so broad, scattered through the writings of Geoffroy Saint-Hilaire in a bold form, received in his conversation a turn so new and so piquant that it inevitably took hold of the thought of any one of liberal mind who tried to trace them to their source.

And when the illuminations of scientific thought were joined to the lively memories of the immortal campaign of Egypt, of which he had been one of the scientific heroes, these recitals, imprinted with the splendors of the Orient, animated by a true poesy, showed by their profound charm with what religious and deeply held sentiments that ardent soul had immersed itself in the ancient Egyptian wisdom. How much livelier yet did this impression become when, struck by illness, his body condemned to an almost absolute repose, his eyes affected by an incurable blindness, he had known how to give to his thought a calm that he abandoned only on rare and solemn occasions, so as to show only that, until his last breath, he would remain faithful to the opinions that had consumed his life! It was scarcely some months ago, last November, that he elaborated to his friend M. Serres views that he asked him to verify with his scalpel, without doubting for a moment that they would come to be put in evidence by his illustrious disciple.

This completes the task of the dean of the Faculty of Sciences. Now let it be permitted to a friend of twenty years to add that, in the double context of satisfactions of the intelligence and of inner happiness, the long sufferings of Geoffroy Saint-Hilaire were alleviated by all the feelings that could touch the heart of men. A family full of respectful veneration lavished on him the most assiduous tenderness; a soul devoted as a son, as the best of sons, surrounded him with the care of a medicine both prudent and learned that seemed for a long time to withhold his prey from an astonished death.

His ardent and tender soul felt so keenly the whole price of the devotions of which he was the object! On his bed of pain, all his words breathed benevolence and inner satisfaction. His hands always searched for his nearest and dearest, his friends, to thank them, to bless them. Calm and smiling, his soul sank without perturbation, withdrew into a conscience without a stain.

For the last two months, the intensity of his illness had doubled; his son, snatched from the duties of the general inspectorship, had had to return to the bed of a dying father; the fatal instant seemed near, inevitable; and yet affection knew how to postpone still further the blow that would strike at one and the same time his family, the Academy, the University, the Museum—which would strike science and, shall we say, France herself in one of her purest glories.

May our words full of our regrets, may our thoughts of sorrow, veneration, and respects, may our prayers at least contribute to bringing some comfort to the soul of the accomplished wife who laments him, of the tender and devoted daughter who made herself his Antigone, of his adopted daughter, and of the son in whom his name is continued, covered with a new renown.

Farewell, Geoffroy Saint-Hilaire, in the name of your colleagues, whose heart will always keep your memory, in the name of your friends engulfed in grief, in the name of your disciples, who leave with regret your mortal remains, but who will always remain full of the spirit with which you have inspired science! Farewell, Geoffroy Saint-Hilaire, farewell!

ORATION OF M. SERRES, MEMBER OF THE ACADEMY OF SCIENCES

Gentlemen:

When a man of genius disappears from the scene of the world, society at once turns its attention to the works that have marked his passage.

The life of the mind is what concerns it, what interests it above all; for society has the innate sense that his productions, more than others, are the ineradicable signs of the greatness of a nation.

France in particular is conspicuous among modern nations for its cult of intelligence, and this character is perhaps most eminent in distinguishing the gallic race from other human races.

Thus, gentlemen, in the religious feeling that brings us together beside the tomb of M. Geoffroy Saint-Hilaire, there is something more than sorrow. There is joined to it that sense of public grief that followed the loss of Lavoisier, of Lagrange, of Laplace, of Chaptal, of Cuvier, of Poisson, whose names and whose memories are so dear to the sciences and to France.

One more name is added to these great names; and like the majority of them, it belongs to the era of the moral renovation of man which began in 1789, which continues without interruption, and in which the sciences play so active a part.

The life of scholars thus finds itself intimately linked to the very life of the nation.

A good act, one of those acts that disclose a whole human life, was the occasional cause of his entry into the Botanic Garden.

Haüy, the creator of modern mineralogy, menaced by the terrible days of September 1792, was saved by Geoffroy.

Geoffroy, sponsored by Haüy, was named subcurator and demonstrator in the cabinet of natural history of the Museum. It was in these modest and learned functions that the Convention's decree of June 10, 1793, found our colleague, to make of him a professor of zoology.

The Convention, which had destroyed everything, found itself in fact under the obligation of rebuilding everything.

Among its creations, there is none that is more deeply imprinted with the greatness of that epoch than that of the Museum of Natural History. There is none which, freed of every political preoccupation of the time, strides toward its goal with more perseverance and elevation without ever straying from its constitutive foundations.

It can well be understood how this decree was administered, how the programs of the twelve chairs were outlined with the high philosophical views that epitomize the eighteenth century. But that at the same moment, twelve scholars would be found to fill them worthily, and would go to work the very next day, that is a phenomenon that only France can produce!

Geoffroy Saint-Hilaire, then so young, and modest, as he always was, hesitated before the task that had been imposed on him, when these words of Daubenton decided his career: Dare to undertake it, he said to the young naturalist, and bring it about that in twenty years one can say: zoology is a French science!

Geoffroy did undertake it, and from his first work on, it was clear that the predictions of Daubenton would not take long to be accomplished.

That first work was, indeed, only a small work on the classification of mammals, a classification that time has barely modified; but the lofty views that it includes place it beside the *primae lineae* of the physiology of Haller; never, in fact, have the two fundamental principles of the natural sciences, that of analogy and that of differences, been combined with more force, with more depth! In meditating on them, one discovers both the genius from which Geoffroy Saint-Hilaire will proceed, and the genius that will produce Cuvier. For in fact, this little work bore the two names, Geoffroy and Cuvier, planted from their very origin like two scientific branches from a common root; the one fructified the differential principle, and the other, the principle of analogy.

The one creating comparative anatomy, and drawing out of it the paleontological theory sketched by Steno and Leibniz.

The other creating zoology and deducing from it the analogical theory of the creation of the animal kingdom.

Sublime conceptions! Worthily closing the eighteenth century and opening the nineteenth with glory! Setting with a bold hand, on one side, the limits of the descriptive sciences, already so advanced, and marking out, on the other, the general or physiological sciences that are only coming to birth! In this way linking the past to the present to smooth the paths to the future.

Invaluable results of the scientific works and of the long professorial career of M. Geoffroy Saint-Hilaire!

Consult the immense works that he has published on the determination of the species, the genera, and the families of mammals and birds; collect the recollections of his lectures, so lively, so original, so winning; everywhere you will find the same philosophy, and as for that spirit and that philosophy, I define them by the words: *the art of observing in the large.*

It is, then, that art that Geoffroy Saint-Hilaire had inherited from Buffon which gained him his successes, and which broke open for him the new paths that he traced in the zoological and anatomical sciences.

Which made him recognize the arbitrariness contained in the classification based on the immutability of species, whose nature showed him their variability at every step.

Which made him seek in the action of external agents the causes of these variations and the reason for the zoological zones of the globe in which families and genera are circumscribed.

Which made him propose the paths of that parallel classification of animals which his son has so clearly formulated, and which presides over the revolution that is taking place at this moment in all the branches of zoology.

There you have the talent of our illustrious colleague; he shows himself in all his works through that observer's glance which discovers at every moment in the animal kingdom characteristics, analogies, differences that had not yet been perceived.

He shows himself, again, by that remarkable character, not of reasoning with more method, but of finding the very principles on which one reasons; not of considering ideas, but of creating new ones and of multiplying them without cessation through profound meditation.

There we have the source of that freedom, that boldness of thought which constitutes one of the most beautiful traits of his writings, and which makes them imperishable, because they are imprinted with that noble independence from the vulgar ideas that limit the advance of the human spirit.

The science of relations, some glimmers of which shine through in the works of Buffon, had found themselves brought to a stop and stifled, as it were, in the details with which Daubenton had surrounded them.

Geoffroy disengages it, and disengages it by one of those bold moves that create a new era in the natural sciences.

This happened when our anatomist wished to explain the bony head of fishes, and to reduce the pieces that compose it to the known vertebrate type and that of adult man. From the first step, he saw that the bony head of man did not provide him with the number of pieces of which that of fishes is composed.

Then, rejecting the idea of ichthyological pieces that his predecessors had admitted, he conceived the idea of looking for the elements that he found missing in the bony anlagen of which the bony head of the human embryo is composed.

Thus he entered upon an immense career whose first steps were crowned with the happiest successes, and which allowed him to conceive of the solution of a multitude of questions formerly considered insoluble, thanks to the futility of the efforts that had previously been broached.

Again, before Geoffroy Saint-Hilaire, philosophy bowed before the gigantic idea of the preexistence of germs and of their eternal encasement; reason kept ceaselessly turning on itself, and after the most painful efforts it found itself back at the same point where it had started. Wisdom consisted in promenading round that circle.

Geoffroy cut through it in his beautiful theory of analogues and carried in his train all independent physiologists.

It is after him and by following the routes that he had traced that we now no longer believe that the embryo of animals is the miniature of their mature state.

It is after him that science has recognized that before stopping at the permanent forms that characterize them, animals traverse a crown of other intermediate forms that are only transitory.

It is since him and with him that comparative embryogenesis, neglected because it seemed pointless, has at present become one of the fundamental parts of zoogeny, the one that commands, the one that dominates them all.

The creative power that systems represented to us, since the first days of creation, as inert and in chains before his own work, diffuses in the new theories its freedom and its action.

The earth becomes a vast laboratory where a succession of new beings continually develop, following a progressive and ascending path, linked together from the infusorians, nature's point of departure, to the mammals and to man, the final term of her efforts: a thought whose wisdom and profundity paleontology every day confirms!

In short, science reveals that progressive and continuous march of life marked at long intervals by halts that seem to be for nature times of rest. Thus does the whole animal kingdom appear as but a single being that, during its formation, stops in its development, here sooner, there later, and thus determines, at each interruption, the distinctive characters of classes, families, genera, and species.

Sublime truth, whose certainty and grandeur modern research increasingly confirms!

Noble friend! Friend of more than a quarter century! May these feeble words rise up to you! May this homage of a heart whose inviolable attachment and sincerity have long been proved to you sweeten the bitterness of the void that your absence will produce in your patriarchal family! . . . which your absence will produce in the soul of all those who have known you, and in scientific Europe, so attentive to your voice!

Farewell, worthy friend, farewell!

ORATION OF M. EDGAR QUINET

Gentlemen:

After so many eloquent eulogies addressed by colleagues to the illustrious deceased, permit a man who has no other right to mourn him than friendship and the assent of his family to add a last word. M. Geoffroy Saint-Hilaire belongs to us all as a portion of that patrimony of glory that France distributes even to the least among us. It is certain that the history of the revolution and its great Egyptian, Spanish, and Portuguese campaigns would not be complete for us, if we did not at the same time see science, with M. Geoffroy Saint-Hilaire, follow the road of the sword and turn to the profit of civilization the upheavals of war. M. Geoffroy Saint-Hilaire in Egypt, at the pyramids, explains and magnifies the destiny of Napoleon, as Aristotle magnifies Alexander. To understand what France can gather together and perform at one time, there has to be a man who, from 1792 to 1815, to 1830, with remarkable consistency, follows a single thought without ever stopping, in the midst of the uproar of revolutions and of battles. The earth is shaken for more than a half century, governments pass, Napoleon falls, another dynasty appears and disappears, and on that constantly shaken soil, in the sort of siege that France is maintaining against the world, there is a thinker, another Archimedes, whom nothing distracts, whom nothing disconcerts; who, his eyes attached to the creation, seeks its mysteries with serenity, as if he did not belong to the region of tempests. Finally, when France has been materially conquered, the obstinate thought of this great mind invades foreign countries, and the greatest writer of Germany, Goethe, seems to have familiarized himself with all the sciences only to inaugurate and to popularize worthily in the world the wholly French victory of M. Geoffroy Saint-Hilaire.

How did it happen that, with so little love of noise and of glory, this man, entirely withdrawn in science, became popular among us? That is because the idea that he brought to light is, in many respects, the foundation of our epoch. Desire, presentiment, necessity of a vast unity: that is what preoccupies the world. M. Geoffroy Saint-Hilaire, truly inspired precursor, established in nature and in science that harmonious principle that we are still seeking in the civil, political, and religious world. That is where the work of that creative mind is linked to the present work of the whole human race; and since he was the first to arrive at that foundation of unity that everyone is seeking along all possible paths, without dreaming of it he has involved everyone in the interests of his fame. We would not be able to follow each of his steps: our ignorance, our impotence bring us up short; we would say to ourselves: he is ahead of us, he is going where the whole century will finally arrive; we would proceed with an assured confidence toward the future, knowing that he already possessed it in the order of science and of nature.

At the same time that science, for him, was always creative, it had I know not what antique and religious character. What persevering enthusiasm, in a time when it is claimed that such enthusiasm no longer exists! What grandeur! What natural amplitude of conception! What patriarchal simplicity! What verve! What inner ecstasy of the man who passes his life in discovering and creating! He is of the family of the Archimedes and the Keplers! He has been accused of being a poet. Yes, without doubt, he was like those great men, through a more abrupt, more imperious, more divinatory presentiment of the exact truth.

After we had received so much enlightenment from that mind in its power, we had still to learn from him, for the last six years, how one ought to die. He went blind like Galileo, but his serenity was never for a moment troubled: he still smiled at the marvels of earth and heaven, which he saw, understood, discovered with the eyes of the mind. One felt in that incredible peace a man who was well aware of the laws and the hidden plan of the creator. He had been initiated into the secret works of Providence, and from that spectacle he had derived the serenity of the just. What could be more sublime than that death of a genius who, thus directed and guided, is the very sanctification of intelligence? Smiling, he approaches truth beyond the veil; at the end, he descends here, fearing nothing, in eternal science.

Where is the one among us, where is the sovereign who would not desire a similar end? And may these words resound to the depths of that empty house, yesterday still filled so completely by the great deceased, whose widow and inconsolable daughter listen to hear the last sound beyond that abyss. They kept him for us for six years, gentlemen, beyond the term marked by nature—those pious hands, which left him during all that time neither day nor night. Tasking that marvel of conjugal and filial piety, he—that just man—said: "I am almost happy to be blind." That they may be requited, those noble women, by the double immortality of him whom they mourn, the much the more so since the son and brother who remains for them recalls the spouse and father who is no more. Among all the families who have brought here their nearest and dearest, how few have obtained what no death can remove! They have almost all left empty-handed, and without present consolation; for you, you take with you, on the contrary, with the glory of the name you bear, a visible immortality, a permanent sign of what our eyes can no longer discern.

M. Geoffroy Saint-Hilaire accompanied our armies in their triumphal course. Is it only mere chance that finds him resting at this moment just beside his friend General Foy? Who among us does not remember that session of one of our academies in which M. Cuvier recounted how M. Geoffroy Saint-Hilaire's devotion has saved our great Haüy from the massacre of September 2? The whole assembly applauds; a man breaks through the crowd, throws himself into the arms of Geoffroy Saint-Hilaire and says to him: "Dear friend, heart, soul, genius, you have everything for you!" That man was General Foy. He was waiting for someone here; the warrior and the scholar had to be once more reunited. Now these two brothers in glory touch one another here in death! Farewell, doubly immortal spirit! You who were so indulgent on earth, do not despise my tribute at this moment. I bid you farewell in the name of all those to whom you have opened a career! Assist me by your light and by your virtue! The best thing in my life will always be to have obtained your friendship.

7

THE DEBATE
A CENTURY AND A HALF LATER

How many commentaries there have been on that controversy! It is true that the two protagonists are fascinating personalities, monuments in the history of biology. The epoch itself is also interesting, as much from a political and social point of view; one has only to read—and to reread—François-René de Chateaubriand (1768–1848) to convince oneself of that.[1] But there is more. In distinction from other famous quarrels, it was not clear until very recently that one could name a "winner," even when using, in anachronistic fashion, the later results of evolutionary biology.[2] With the passage of time, nothing would really appear to be clarified, since serious errors were to be found in both zoologists, errors that placed them back to back.

Certainly, Baron Cuvier's mistake was a major one: his denial of evolution will be an eternal reproach to him. His taste for honors and for power makes him a not very likeable personality.[3] Nevertheless, it will not do to reject altogether the whole of his work, and to fail to recognize in him a particularly acute analytical mind. He was, without any question, an anatomist and a paleontologist of genius who advanced comparative biology by a giant step. Nor can one deny that Cuvier had good reasons to oppose Geoffroy Saint-Hilaire. The existence of a unique plan of organization did not have enough support from a scientific point of view; the insects considered as dermo-vertebrates, the existence of a diaphragm in the cuttlefish: those were so many alarming propositions, long since refuted, and which now make us smile.

If Geoffroy Saint-Hilaire's transformism now seems laudable, still we cannot pass over in silence the archaisms of his thought. Despite his attempts in other directions, he never shook off the image of the Scale of Beings, which leads us to arrange animals according to a simple *scala naturae,* according to their "degree of perfection," a Platonistic vision that would lead Richard Owen to the simplistic concept of the archetype.[4] But for Cuvier, it is equally true that one cannot deny the importance of his results in comparative anatomy and embryology, in teratology and evolutionary paleontology. Since the protagonists both made bad mistakes, it is easy for anyone who wants to defend his favorite to prove that the other was wrong. But since this operation can be carried out from either side, the analysis of the controversy soon becomes circular.

It is also advisable to reject certain received ideas. Thus the controversy is sometimes presented as a new version of the combat of David against Goliath, the little zoologist against the Pope of Science, who amasses glory and power. That is to forget that Geoffroy Saint-Hilaire, professor at the Museum and at the Sorbonne, protégé of Bonaparte and friend of numerous high functionaries, was himself a man of great influence, an influence that did not stop increasing from 1818 onward. But rubbernecks love caricatures, and we always like to see the person of power—or of the most power—brought to his knees.

Oddly enough, a century and a half later this controversy still arouses passions. Thus René Thom declares, in 1975: *"I would not be surprised if in the future more justice was rendered to the considerations of Geoffroy Saint-Hilaire, even if at present they still appear delirious."*[5] Perhaps, beyond the sphere of science, a debate of this kind touches certain sensitive fibers of the human soul, bearing on the domain of the irrational, if not the mythical. We have seen that the comprehensive vision of Geoffroy Saint-Hilaire also touched on the status of man; his animal origin is demonstrated; by analogy, a law of "progress" seems to be applicable to humanity. All this is full of dreams, to be sure, but also of hope.

What happens to Geoffroy Saint-Hilaire's ideas after his death? Darwinism, after it has triumphed from a scientific point of view, takes over everywhere: social Darwinism prevails, especially in the English-speaking countries.[6] Conceptions of embryology and evolution, combated in France by Cuvier's students, are reinterpreted in England in this new evolutionary framework. The phylogenetic position of man among the primates is reexamined in a manner more deeply developed by Darwin himself.[7] Finally, the naturalist's humanistic vision finds itself relegated to the category of nineteenth-century idealism.[8] Thus the French epoch marked by the passing splendor of the Museum of Natural History appears to belong to a time that is definitively over, and the ideas of Geoffroy Saint-Hilaire thus seem to be able to interest only a few historians of science.

It is at that point that the results of developmental genetics have recently arrived on the scene like a thunderbolt. The discovery of genes which intervene precociously in the development of the embryo, and which have been playing an important part ever since the establishment of the body plans of animals, has opened the way to a new type of relationship between embryology and evolution, that is to say, between ontogeny and phylogeny. Body plan, embryology, evolution . . . Would we not have, this time, the necessary givens to propose at last a reasonable conclusion to the controversy between Cuvier and Geoffroy Saint-Hilaire?

Let us pause an instant at the notion of body plan. If we consider animals in their totality, what is there in common between a fly and a mouse, an earthworm and a sea urchin? A priori, nothing, if not some functional possibilities (displacement, nutrition, sensibility) and the cell. All these organisms are made of cells, but they are arranged in different fashions.[9] If their constituents are always the same, their plans are globally different. It is then that there appears in all its importance a key result of nineteenth-century biology: the plans existing in the animal world are of a limited number.

It is not very difficult to find a single plan among the vertebrates: as early as 1555, Pierre Belon (1517–1564) divulged it by comparing the skeletons of a man and a bird.[10] Later, a generalization was made for the tetrapods (four-footed vertebrates), then, with further abstraction, for the fishes. The body plan of the vertebrates is characterized essentially by the succession, from back to belly, of a central nervous system, a vertebral column, and a digestive tube whose anterior part, or pharynx, is pierced by gill slits.

In the same way, we can easily find an identical organization among the insects, by comparing, for example, a dragonfly, a cockchafer, a butterfly, and a bee. Zoologists generalize for the whole of the arthropods, adding the crustaceans, the myriapods (centipedes), the arachnids (spiders and scorpions). In the arthropods, unlike the vertebrates, the central nervous system is ventral. One can proceed in the same way, separately, for the mollusks and for the annulated worms (annelids).

Thus, at the dawn of the twentieth century, it was acknowledged that each great group—subkingdom or phylum—is in fact characterized by a body plan.[11] This looks like a "tool" allowing us to bring together animals that are, as a whole, organized in a similar manner. Within the interior of each subkingdom (chordates, arthropods, mollusks, annelids), we can easily inventory and classify animals that present an identical basic structure. We compare what is comparable, and we can describe the diversity after we have established the unity.

But how to classify the subkingdoms? The body plans are apparently irreconcilable. We do indeed possess some criteria for comparative embryology. One of the most important is the number of germ layers, which are at the origin, in the young embryo, of the various organs of the animal. There are only two of these—the external germ layer or ectoderm and the internal germ layer of endoderm—in certain animals, which, like the medusae and the sea anemones, present only one external layer of cells and a digestive tube.[12] In others—the great majority—we find three. The latter possess in addition an intermediate germ layer or mesoderm, which is found, for example in the vertebrates, at the origin of the skeleton, of the muscles, of the blood, of the heart, of the kidneys, and also of the gonads.[13] In the vertebrates, the ectoderm gives rise to the epidermis and the nervous system, while the endoderm forms the digestive tube and the glands annexed to it.

Thus it was logical to consider a certain hierarchy among animals; before having three germ layers, it was first possible to have two, supposing that in this domain the simple precedes the complex. We can once more apply the rule that holds that to an important structural element there corresponds an historical unity. Thus all animals have, at a given moment, a distant common ancestor with two germ layers. Then, for the more complex among them, we assume the existence of a more recent common ancestor which acquired the third germ layer. Such embryological considerations—as well as others—are essential; but this concerns only a few characters, in any case not a sufficient number to group together the whole of the subkingdoms, or to produce a unity among all those different body plans.

In short, we found ourselves, even a decade ago, in a rather unsatisfactory situation. At the cellular level, we detected a profound unity, valid for the whole of life. But when we confined ourselves to the animals, such a unity was not discernable except at the interior of each subkingdom. If each of them, characterized by a body plan, seemed to represent an entity in itself, the bond between them seemed very difficult to establish. Paradoxically, the animal kingdom, so familiar in its entirety, did not allow us to discern any unity at the morphological level and thus seemed opaque when taken as a whole.

Thus, although for some years the unity of the whole of life already appeared to be well supported biologically, up to the present this was apparent only at the cellular or subcellular level. It is indeed clear that all cells, whatever their type, present the same dynamic: the long molecule of DNA is everywhere the basis of the genetic information, and it determines—or codes for—the structure of proteins, molecules the majority of which unite, through their catalytic power, in the realization of the whole of the chemical reactions of living matter. From an evolutionary point of view, such a unity of cellular function is interpreted as the sign of the existence of a distant common ancestor, and thus of a unity of origin.

This beautiful unity of the living world even goes so far as to include the bacteria. But, concerning as it does only the molecular and cellular levels, it does not explain the essential differences that are apparent to direct perception. When we take an interest in organisms, it is not the unity that strikes us, but rather their extraordinary diversity, the source of biodiversity, following a concept henceforth known to the general public.[14]

It is in this context that the recent and highly significant discovery of developmental genes intervenes in a decisive manner. In the past fifteen years, biology has been going fast, very fast. The advances made by molecular techniques now allow us to reach directly the heart of the genome and, very recently, those particular genes that control the development of the embryo. What a revolution! Instead of remaining at the stage of description, we have access to bodies of genetic information that punctuate and measure rhythmically the ballet of the cells that lead, after fertilization, to the construction of an animal.

It all began around the 1950s, with the study, first by classical genetics and then by molecular genetics, of particular mutants, the homeotic mutants. Such mutants had been known for a long time, for they were first described by William Bateson (1861–1926)[15] some years before the rediscovery of Mendel's laws and the foundations of chromosomal genetics.[16] Bateson coined the term *homeosis* to designate the presence in animals or plants of well-formed organs located in abnormal places. Thus in insects, the mutant called "antennapedia" has legs on the head instead of antennae. As to the mutant "bithorax" in drosophila—a dipteran insect, thus possessing a single pair of wings—, it gives the animal two pairs of wings. Everything happens as if it were exclusively the "address," and not the construction, of the supernumerary organs—legs in the one case, wings in the other—that was wrongly determined. These mutations concern particular genes, called homeotic; they normally intervene in the formation of the embryo and participate in the structuring of the plan of organization of the animal.[17] Later, a detailed study allowed geneticists to show that, in the cascade of developmental genes that intervene, the homeotic genes are the last to develop. Further, it is known that the homeotic genes are groups in a totality called the Hox complex. The genes of this complex control one of the great axes of development: the anterior-posterior axis, which determines the harmonious sequence of all the organs, from the head to the posterior extremity of the body. They obey the rule of colinearity, that is to say, their position in the complex corresponds to the anterior-posterior polarity of their expression in the larva.

At the turn of the 1980s several teams discovered that these homeotic genes all have a common sequence: in other words, a fragment characterized by the same series of nucleotides. This sequence was called the homeobox (hence the name given to the Hox complex).[19] The fact that these genes occur in such a sequence proves that they are homologous, that is to say, that they derive from an ancestral gene that has been duplicated several times in the course of evolution to form a family of several genes (a multigene family).

Once the existence of this homeobox in drosophila was known, its presence was looked for in other animals, in other arthropods, but also in vertebrates. The results exceeded all expectation. The comparison of animals as different a priori as drosophila, mouse, the zebra fish *Brachydanio rerio,* or the nematode *Caenorhabditis elegans,* a little roundworm that contains only about a thousand cells, has led to a surprising discovery: in all these organisms, it is the same genes that control their anterior-posterior organization. This result was not easily achieved. When these genes were first described in drosophila, it was

HOW A FLY IS BUILT

A drosophila consists of sixteen segments. The first five correspond to the head, the next three to the thorax, the last eight form the abdomen. The first thoracic segment carries only one pair of legs, the second carries a pair of wings and a pair of legs, the third a pair of balancers (organs serving for equilibration during flight) and a pair of legs. The eight segments of the abdomen are deprived of legs and all the other appendices, with the exception of the last, which carries the copulatory organs.

The embryonic development of drosophila obeys a series of orders that are now well known. After the establishment of a double anterior-posterior and dorsiventral polarity, the embryo, which appears in the form of an elongated ovoid, is divided into several major regions, corresponding roughly to one in front, the head, then the thorax, and the abdomen in the rear. Each of these regions is again segmented, then each of the segments acquires an anterior-posterior polarity, so as finally to have its own characteristics specified, like the appendices (antennae, jaws, legs, etc.).

Each of these operations depends on particular genes. Maternal genes for polarity chiefly ensure the acquisition of polarities. Then genes of segmentation, which can be divided into three groups, determine the way the embryo is structured. First, the gap genes specify the division of the embryo into large zones; then the pair-rule genes determine the segmentation of the organism; finally, the segment-polarity genes produce the anterior-posterior polarity of the segments. Thus there are three levels of control for a correct segmentation of the organism. It is only after the expression of the totality of these genes that the homeotic genes begin to act, intervening to give each segment its particular anatomy.[18]

quickly understood that we had laid hands on major molecular instrumentalities participating in its construction. But obviously, it was first thought that they specified the body plan of that particular animal. Then it was easy to imagine that such genes could also be found in other insects, even in other animals in the same phylum, like the crustaceans or the arachnids, which would thus have an identical body plan. But very quickly these genes were identified in mammals, in which they play a very similar role. Suddenly a factor arose that had been unsuspected until then: these homeotic genes establish the anterior-posterior polarity of the body, that is to say, they determine the place of the future head, of the future tail, of the major divisions like the thorax and the abdomen, and they do this, no matter what the animal's body plan. The molecular

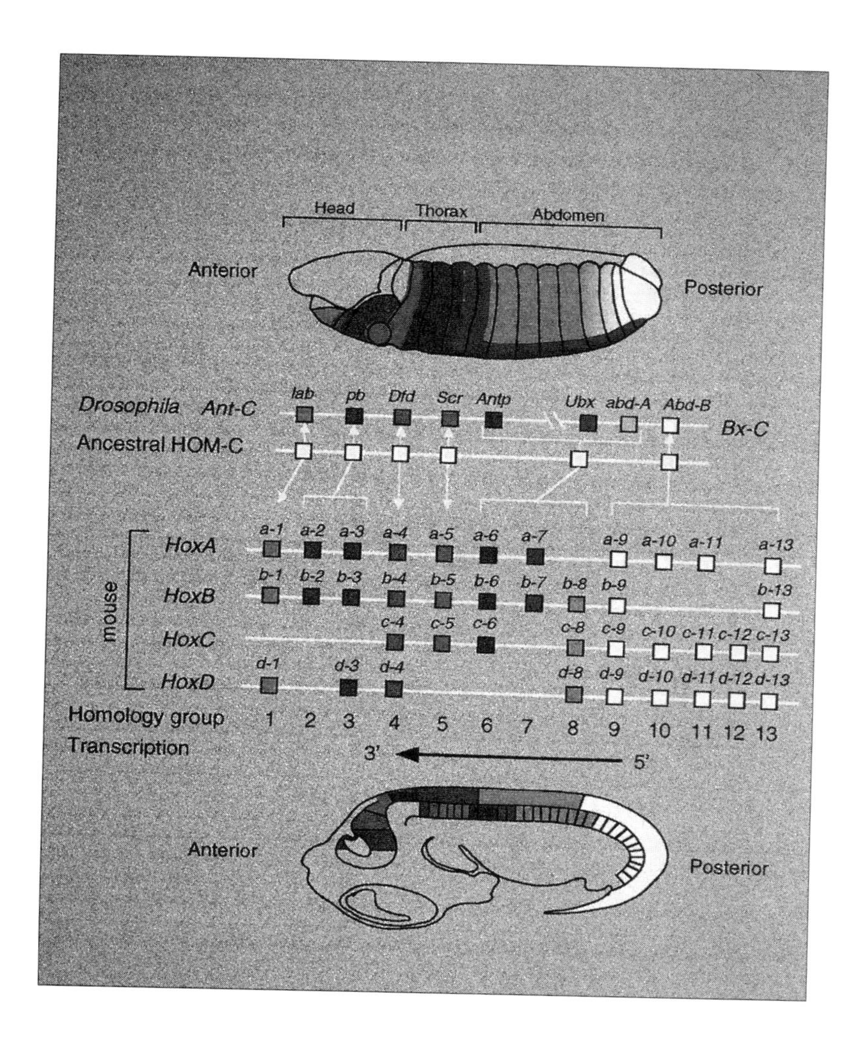

system that fashions the composition of the organism thus seems to be unique, and not to be dependent on that organism's body plan. The difference between these two concepts is subtle, but of major importance. Species as different as an insect and a vertebrate share not only homologous genes, but, equally, homologous complexes, all of which obey the rule of colinearity.

Such similarities can be explained only if vertebrates and insects have inherited from a common ancestor a molecular representation of the "front," "middle," and "rear" of the animal. Current knowledge of the homeotic complexes is such that we have come to postulate the structure of the complex in the hypothetical common ancestor of vertebrates and arthropods, which would have consisted of seven genes. Thus everything leads us to believe that arthropods and vertebrates have inherited this same genic organization from a common ancestor, necessarily exceedingly ancient. According to paleontologists, that common ancestor would have lived at the beginning of the Paleozoic Era, about 540,000,000 years ago.[20]

The existence of a common ancestral complex for the insects and the vertebrates has clearly led biologists to look for this in other animals, including those that present an apparently simpler organization.[21] To sum up: to date, homeotic complexes have been found with certainty in the chordates (and thus the vertebrates), the arthropods (crustaceans and insects), and the nematodes; other partial results allow us to think that all triploblastic animals (animals with three layers of cells) possess such a complex. On the other hand, diploblasts like the cnidarians (sea anemones, corals, and medusae) seem to have homeobox genes, but not complexes. Finally, the sponges seem at present to resist all investigation.

So here we have at the embryological level an important common point between arthropods and vertebrates . . . And we begin to rediscover Geoffroy Saint-Hilaire. In fact, the whole of these results suggests a great unity in the animal kingdom, a more profound unity than that envisaged by comparative anatomy, since it concerns organisms with different body plans. In 1993, this discovery led a group of English embryologists to propose a new definition of an animal: an animal would be an organism possessing a particular spatial organization resulting from the expression of the Hox complex genes, an organization called the zootype.[22] A zootype—a badly chosen substantive—is not an organizational archetype in the sense of Owen, but an ordered set of items of information specifying the sequence of the parts of the organism from anterior to posterior.[23]

Let us consider what such a concept implies. First point: triploblastic animals can present different body plans—that of an insect is not the same as that of a vertebrate, each of them having the principal characteristic of a particular group. Second point: such organisms, presenting different plans, nevertheless possess the same battery of genes, which are expressed everywhere in the same way—following the rule of colinearity, with the function of determining the anterior-posterior direction of the embryo.

Would not this be a modern vision of "organic composition," a term chosen by Geoffroy Saint-Hilaire, after some hesitation, to designate his law of unity, if we attribute to the term "composition" that of a pictorial and rhetorical metaphor?[24] Let us not forget Cuvier's remark as to Geoffroy Saint-Hilaire's comparisons of insects and vertebrates: "*Nothing in common, absolutely nothing, between the insects and the vertebrates; at the very most one single point, animality.*" But it is precisely animality that is here in question. And in the concept of zootype do we not recover the notion of virtuality—or potentiality—dear to Geoffroy Saint-Hilaire, applied, though awkwardly, during the controversy on the hyoid bone? It seems in fact that the existence of this zootype would be one of the factors that would have permitted the astonishing diversification at the moment of the explosion of life in the Cambrian, at the beginning of the Paleozoic Era.[25]

Diversity of body plans or unity of composition? Both, obviously; the antinomy is only apparent. An insect and a vertebrate do not have the same body plan, but they do present the same system of anterior-posterior ordering. The diversity entails the unity, and the unity authorizes the diversity. In a synthetic fashion, we could consider that the body plan is a structural, static concept; it becomes clear when embryogenesis is terminated (situation of the central nervous system, of the digestive tube . . .). Composition is an essentially dynamic concept and denotes a precocious operation which, in embryological time, corresponds to a sketch preparing the execution of the body plan. Such a sketch is effected in the same way, by the same molecular instrumentalities, for body plans that are a priori totally different.

Thus we can consider the embryology of an animal from two points of view that are extremely different, but complementary. Either one focuses, for each group, on the body plan that characterizes it, and so points to the diversity that deeply divides the animal world. Or one looks at the dynamic of the initiation of the anterior-posterior polarity, which allows the determination from front to back of the arrangement of the organs, their ordering, their composition, that is to say, what most fundamentally permits the implementation of the body plan. One is then assenting, through this apparently common dynamic, to an astonishing unity of the animal world. Thus we rediscover, at the level of the development of the organism, a unity which, because it transcends the various body plans, will permit us to regroup the different subphyla with one another; now, the animal world is one.

But what is most fascinating is still to come. In fact, the zootype is not concerned only with anterior-posterior organization. Everything was overturned when, in the 1990s, people began to understand the dorsal-ventral organization (upper face and lower face) of the triploblasts and to see that it is likewise guaranteed genetically in the same way in all animals. In drosophila, two major genes have been discovered, which intervene in antagonistic directions: the gene *sog*, which determines the ventral part of the embryo, and the gene *dpp*, which, for its part, is expressed dorsally. Their complementarity allows the realization of a harmonious embryo. In a vertebrate, in the case of the toad xenopus, we find a similar gambit: the gene *BMP-4* is expressed ventrally, while the gene *chordin* determines the back, where the spinal cord is found.[26]

The surprise was great when the sequences of genes were determined. The gene *sog* is homologous to the gene *chordin*, while the gene *dpp* is homologous to *BMP-4*. In other words, the insect's gene with ventral expression is the same as that with dorsal expression in the vertebrate, and vice versa! Now let us not forget that the nervous system is dorsal in the vertebrates and ventral in the insects. Thus if the anterior-posterior signal is identical in an arthropod and a vertebrate, the dorsal-ventral signal is inverted. It would suffice to turn one of the animals upside down and everything would agree. That is what Geoffroy Saint-Hilaire had done with his lobster. It is easy to understand why the geneticists who have obtained these results are happy to cite Geoffroy Saint-Hilaire!

Naturally, the actual explanation is foreign to Geoffroy Saint-Hilaire's argument. In fact, it is the situation of the mouth that is the key element; that is what determines which aspect is called "oral" and which is called the "abdomen." But embryology shows that as distinct from that of the insects, the mouth of vertebrates is a structure that appears secondarily in the course of embryonic development, after the anus.[27] The true novelty in the vertebrates is the situation of the mouth. But the situation of the central nervous system is determined by the same gene, *sog* or *chordin*, in the insects and in the vertebrates. Only it is called "abdomen" in the insect and "back" in the vertebrate. In fact, the difficulty comes from the mouth—and from words! If the surfaces had been distinguished not by the position of the mouth, but by that of the nervous system, everything would be homogeneous.

Was Geoffroy Saint-Hilaire Right?

Let us transport ourselves in thought for a moment to the eve of the discovery of homeotic genes, to the moment at which we had not yet any tangible trace of that unity. What to say to those who, before making out the evidence, suspect it, have a hunch about it? No doubt the comment would be, "You are dreaming, dear colleague!"

That is the reply that Etienne Geoffroy Saint-Hilaire heard throughout his whole career. Material, tangible facts were constantly demanded of the man who untiringly affirmed, for theoretical reasons, that unity of organic composition of which he had had a premonition since 1796. There were facts, but facts furnished by Georges Cuvier, who argued inexorably for the diversity of plans of organization. And nevertheless, by a strange destiny, it is Etienne Geoffroy Saint-Hilaire who, after a very long absence, returns to the first level of the international scientific scene.

Thus in 1988, Stephen Jay Gould, professor at Harvard and highly gifted popularizer, published in the *American Naturalist* an article entitled "Geoffroy and the Homeobox."[28] As we have seen, the homeobox is one of the most elegant indications of that unity at the molecular level. The subtitle of the article hit the nail on the head: "The art of finding timeless essences in apparent trifles is the kind of perception that we call genius." There followed a parallel presentation of the thought of Geoffroy Saint-Hilaire and of the principal findings of molecular genetics having to do with the homeotic genes that determine the anterior-posterior organization of the embryo.

In exemplary manner, there are quantitative facts that speak for themselves. Developmental biology is a discipline that is progressing very rapidly at present; that is why the bibliography of scientific articles on the subject rarely contain references that are more than five years old. That is why, in a recent article by two English authors with the title "Revolving Vertebrates," published in *Current Biology* in July 1995—a scientific review for specialists—there is a revealing anachronism.[29] The bibliography includes sixteen references to articles published between 1991 and 1995, and one to an article from 1822! Had nothing happened between 1822 and 1991? The subtitle of the article is once again explicit: "An old idea about the relationship between arthropod and vertebrate body plans has been given new life by studies of the signalling genes controlling dorsal and ventral development in Drosophila and Xenopus." That "old idea," published in 1822, is of course that of Geoffroy Saint-Hilaire.

> Functional studies seem now to confirm, as first suggested by E. Geoffroy Saint-Hilaire in 1822, that there was an inversion of the dorsoventral axis during animal evolution. A conserved system of extracellular signals provides positional information for the allocation of embryonic cells to specific tissue types both in *Drosophila* and vertebrates; the ventral region of *Drosophila* is homologous to the dorsal side of the vertebrate. Developmental studies are now revealing some of the characteristics of the ancestral animal that gave rise to the arthropod and mammalian lineages, for which we propose the name *Urbilateria*.
>
> Excerpted from E. M. De Robertis and Yoshiki Sasai, "A Common Plan for Dorsoventral Patterning in Bilateria," *Nature* 380 (1996): 37–40.

Finally, in March 1996, in the prestigious review *Nature* the French naturalist was again honored, in an article that develops a new series of arguments stemming from molecular biology. The first phrase of the heading is unambiguously on the role of Geoffroy Saint-Hilaire, who, on this occasion, is explicitly named: "Functional studies seem now to confirm, as first suggested by E. Geoffroy Saint-Hilaire in 1822, that there was an inversion of the dorsoventral axis during animal evolution."[30]

Thus, blow by blow, Geoffroy Saint-Hilaire's visions on the anterior-posterior, but also on the dorsoventral, organization of animals have been rehabilitated. What an acknowledgment, more than a century and a half later! There is naturally no question of wanting to say that Geoffroy Saint-Hilaire had suspected the existence of developmental genes. At the time, the very existence of the genes as such had not yet been suspected. What it is appropriate to emphasize is simply that ideas developed at the beginning of the nineteenth century, and so strongly attacked at the time, now find themselves confirmed by modern biology. Geoffroy Saint-Hilaire was right intuitively, but let us repeat once more: there is no question of passing over in silence his errors and inaccuracies or of criticizing Cuvier's work. Let us simply agree that he acted like a visionary of genius. If Cuvier was wrong to be right, Geoffroy Saint-Hilaire was right to be wrong. A century and a half later, which is the more enviable situation?

NOTES

Introduction

1. Comparative anatomy and paleontology were founded chiefly by Cuvier and Geoffroy Saint-Hilaire.
2. The theory of evolution was taught by Lamarck in 1802 (opening lecture of the course, 27 Floréal, year X), then published in 1809 in his *Philosophie zoologique.*
3. Cholesterol was discovered by Eugène Chevreul (1786–1889), a chemist specializing in fatty substances. This molecule was baptized in 1815 by Chevreul as "cholesterine," then called "cholesterol" by Marcelin Berthollet (1827–1907) in 1859 because of its alcoholic function.
4. Henri Becquerel (1852–1908) discovered radioactivity in 1896 while carrying out research on the X-rays that had recently been brought to light by Wilhelm Conrad Röntgen (1845–1923).
5. The Great Gallery of Evolution corresponds to the former Gallery of Zoology, on the rue Geoffroy Saint-Hilaire. It was opened by the president of the republic in 1994. For its history, see D. Bezombes, *La Grande Galerie du Muséum National d'Histoire Naturelle* (Paris: Le Moniteur, 1994).
6. Following the title of the excellent book by François Dagognet, *Le Catalogue de la vie* (Paris: Presses Universitaires de France, 1970).
7. François Magendie (1782–1855), a physiologist specializing in the nervous system, was one of Cuvier's protégés.
8. Claude Bernard (1813–1878) advocated the study of the function of organs. From this point of view he is a descendant of Cuvier. See M. D. Grmek, *Claude Bernard et la méthode expérimentale* (Paris: Payot, 1991).
9. Louis Pasteur (1822–1895) always carried out a functional study of whatever he was investigating, whether it was crystals or microorganisms. See R. Vallery-Radot, *La Vie de Pasteur* (Paris: Flammarion, 1900).
10. After the publication of Darwin's *Origin of Species* in 1859, certain professors at the

Museum played a role that proved prejudicial to the development of evolutionary biology in France. Etienne Serres and Armand de Quatrefages could be mentioned in this connection.

11. The first chair in genetics at the Sorbonne was not created until after the Second World War, and this despite the importance of a figure like Lucien Cuénot (1866–1951). In reality the first chair of genetics in France was created at the National Institute of Agronomy. In fact, Félicien Boeuf was recruited as professor for Ducomet's chair, which was announced with the title "chair of genetics" (Georges Valdeyron, *Souvenirs,* unpublished; Jean Gayon, personal communication).

12. In a large area around the Museum, we find the names of almost all the professors who were of any importance. Nevertheless, there is one exception. To find the rue Lamarck you have to go to the 18th arrondissement. The rue Darwin is in that same arrondissement. See A. Alter and P. Testard-Vaillant, *Guide du Paris savant* (Paris: Belin, 1997).

13. Ernst Mayr and Stephen Jay Gould discuss him at length. France is a strange country, which seems to be ashamed of its great men. While in England the whole of Darwin's work is available—including the entirety of his correspondence—nothing of the sort exists for Cuvier, Lamarck, or Geoffroy Saint-Hilaire. See S. J. Gould, *Ontogeny and Phylogeny* (Cambridge: Harvard University Press, 1977); E. Mayr, *The Growth of Biological Thought* (Cambridge: Harvard University Press, 1982).

14. The memoir of 1796 had to do with the makis, which are lemurian primates of Madagascar. See E. Geoffroy Saint-Hilaire, "Mémoire sur les rapports naturels des Makis Lémur L. et description d'une espèce nouvelle de Mammifère," *Magaz. encycl.* 2 (1796): 20–50.

Chapter One

1. In a text of 1835, "Etudes progressives d'un naturaliste pendant les années 1834 et 1835," forming an appendix to his publications in the forty-two volumes of the *Mémoires et annales du Muséum d'Histoire Naturelle* (Paris: Roret, 1835), we find a touching and astonishing note in which Geoffroy Saint-Hilaire reports the last conversation that Bonaparte had in Cairo before his departure for France, and tries to make Bonaparte appear as a visionary in science. This text must be read as evidence for the admiration that Geoffroy Saint-Hilaire felt for the warrior, even after the Restoration:

> The conversation was serious and remarkable for its uniform tendency; here is a summary of it, insofar as I remember it.
>
> Bonaparte spoke as follows: "We have leisure, Monge; let us pass the time in philosophizing, and to give us a subject, I shall tell you about the thoughts I had in my youth. The profession of arms became my occupation; it was not my choice, and I found myself caught up by circum-

stances. When I was young I had taken it into my head to become a discoverer, a Newton." "What are you saying, General?" Monge answered, "it seems you don't know the saying of Lagrange: No one will achieve the glory of Newton, for there was only one world to discover." "Oh! What opposition are you setting up there, M. Monge? Our friend Berthollet, profound in his knowledge of the play of affinities, on the subject of molecules as principles, no, without doubt, Berthollet is not at all of your opinion. Who called attention to the character of intensity and of traction at a very short distance in the actions of particles, of which we are witnesses every day? Monge, has that been found? You, Monge, or your Newton, would you have found it? But, look: would not this be lovelier, greater, but above all more profitable to society, than a philosophical speculation? Newton resolved the problem of movement in the planetary system; that is magnificent for you intellectuals and mathematicians. But, as for me, if I had taught men how the motion operates that is communicated and determined in small bodies, I would have resolved the problem of the life of the universe. And with that accomplished as I supposed, I would have outdone Newton by all the distance that obtains between matter and intelligence. Hence there is nothing at all exact in your saying of Lagrange. The world of details is still to be investigated. So there is that other world, the one that is the most important of all, that I had flattered myself I would discover. Thinking of it always fills me with regret; thinking of it sickens my soul!"

Did Bonaparte really say that? Geoffroy Saint-Hilaire seems very convincing.

2. The two best books on the life of Geoffroy Saint-Hilaire are the biography written by his son Isidore and the book (unfortunately out of print) by Théophile Cahn: I. Geoffroy Saint-Hilaire, *Vie, travaux et doctrine scientifique d'Etienne Geoffroy Saint-Hilaire* (Paris: Bertrand, 1847; reprinted, Brussels: Editions Culture et Civilisation, 1968); Th. Cahn, *La vie et l'oeuvre d'Etienne Geoffroy Saint-Hilaire* (Paris: Presses Universitaires de France, 1962). We may also mention the article "Geoffroy Saint-Hilaire Etienne et Cuvier Georges, Débat" in P. Tort, ed., *Dictionnaire du Darwinisme et de l'Evolution* (Paris: Presses Universitaire de France, 1996), 1867–1883.

   On the controversy, in addition to the works referred to, see T. A. Appel, *The Cuvier-Geoffroy Debate* (Oxford and New York: Oxford University Press, 1987).

3. At the time Daubenton was one of Buffon's closest collaborators, only to quarrel with him later.

4. The decree of July 10, 1793, specified the twelve following chairs:

   Mineralogy: Louis Jean Marie Daubenton (1716–1800)
   Geology: Barthélémy Faujas de Saint-Fons (1741–1819)
   Botany at the Museum: René Louiche Desfontaines (1750–1833)

Botany in the field: Antoine-Laurent de Jussieu (1748–1836)
Cultures: André Thouin (1747–1824)
Zoology (insects, worms, microscopic animals): Jean-Baptiste de Lamarck (1744–1829)
Zoology (vertebrates): Etienne Geoffroy Saint-Hilaire (1772–1844)
Animal anatomy: Jean-Claude Mertrud (1728–1802)
Human anatomy: Antoine Portal (1742–1832)
General chemistry: Antoine François de Fourcroy (1755–1809)
Chemical arts: Antoine Louis Brongniart (1742–1804)
Iconography: Gérard Van Spaendonck (1749–1822).

5. To be exact, the chair assigned to Geoffroy Saint-Hilaire was divided in two upon the return of Etienne de La Ville-sur-Illon, count of Lacépède. Geoffroy Saint-Hilaire then took the mammals and birds, Lacépède the reptiles and fishes. In reality these titles were fictitious enough, for we see Geoffroy Saint-Hilaire publishing extensively on fishes and crocodiles.
6. Certain historians, among them Appel, give as the first holder of the chair Antoine Louis François Mertrud, the son of Jean-Claude. According to Brygoo—whom I follow here—it was Jean-Claude who gave the course in 1794; it is he also whom Haüy lists among the first professor-administrators; it is to his position that Cuvier succeeds. Cf. E. R. Brygoo, *Les Professeurs-administrateurs du Muséum National d'Histoire Naturelle et leurs chaires* (Paris: Aide-Mémoire, Bibliothèque centrale du Muséum National d'Histoire Naturelle, 1990).
7. Geoffroy Saint-Hilaire would retain the management of the menagerie of the Botanic Garden until 1841, except for six months during which Frédéric Cuvier, Georges's younger brother, was in charge. In fact, a chair of comparative physiology, to which the menagerie was attached, was created for Frédéric on December 12, 1837. But he died on July 24, 1838, and Geoffroy Saint-Hilaire reclaimed his own. In the manuscripts preserved at the Academy of Sciences is a letter by Geoffroy Saint-Hilaire protesting against the decision of 1837. This letter illustrates his rough and hot-headed style: "You call me on February 25 to be at your service at a given hour, but do you know if I am well enough for that, at the end of the illness that a blow of your injustices had inflicted on me and to which I was bound to succumb, after you had hurled me out of my social position and the services of my functions because, one of your people told me, you are a Chamber of Deputies on a small scale, and because you have the right to strike outcasts and to treat them like Manuel, who was seized and thrown backward? Manuel did not have my rights and mine were to possess the Menagerie as a thing belonging to me, in view of the fact of its creation. I created it hook, line, and sinker and for ten years, from the year II to the year XII, through incessant labor and innumerable difficulties that were nevertheless surmounted." Cited in Y. Laissus, "Catalogue des manuscrits d'Etienne Geoffroy Saint-Hilaire conservés aux Archives de l'Académie des Sciences de l'Institut de

France, *Histoire et nature* 1 (1973): 71–88. (Jacques Antoine Manuel [1775–1827], who was elected as a deputy in 1818, belonged to the liberal opposition. He was forcibly excluded from the Chamber of Deputies because of his opposition to the Spanish expedition in March 1823.)

Etienne's successor in the direction of the Menagerie would be his son Isidore, who managed it with competence from 1841 to 1861.

8. G. Cuvier and E. Geoffroy Saint-Hilaire, "Sur les espèces d'eléphants," *Bull. Sc. Sté. Philom. Paris, 1791–1796*, 1 (1795): 90; G. Cuvier and E. Geoffroy Saint-Hilaire, "Histoire naturelle des orangs-outangs," *Magaz. encycl.* 3 (1795): 451.
9. In fact the application of the principle of the subordination of characters to the animal kingdom is already adumbrated in a joint text by Cuvier and Geoffroy Saint-Hilaire, published in 1795 ("Mémoire sur une nouvelle division des Mammifères et sur les principes qui doivent servir de base dans cette sorte de travail," *Mag. encycl. du J. des sciences, des lettres et des arts* 2 [1795]: 164). Nevertheless, the foundational work will be Cuvier's alone (*Tableau élémentaire de l'histoire naturelle des animaux* [Paris, 1798; reprinted, Brussels: Editions Culture et Civilisation, 1968]).
10. Why is it Berthollet who was the great organizer of the scientific expedition rather than Monge, who was much more deeply involved politically? To understand this, we have to go back to Bonaparte's first Italian campaign. In fact, in 1796, both Monge and Berthollet, along with four other colleagues, were assigned the delicate mission of going to collect in the newly conquered countries whatever would supply materials for French museums. It was then that Berthollet came to know Bonaparte, who became his friend and would always set great store by his chemist. But in 1798 Monge was still in Italy; thus it was Berthollet who acted in his name. After the fitting out of the expedition, the two scientists embarked on the flagship L'Orient with Bonaparte. Later on it would be Monge who would be president of the Institute of Egypt. See M. Sadoun-Goupil, *Le Chimiste Claude-Louis Berthollet (1748–1822), sa vie, son oeuvre* (Paris: Vrin, 1977).
11. Opinions are divided on the causes of Cuvier's refusal. However, I prefer to follow Bultingaire, who demonstrates that Cuvier was ill at the time. It is worth noting that Cuvier would never be a traveling naturalist. Everything he wanted to observe he had brought to the Museum. See L. Bultingaire, "Iconographie de Georges Cuvier," *Archives du Muséum*, 6th ser., 9 (1932): 1–12.
12. Unlike Geoffroy Saint-Hilaire, Fourier would accept a position as prefect. In April 1802, at the age of thirty-four, he arrived at the prefecture of l'Isère. This would lead him to complain constantly of not being able to devote himself exclusively to science. However, despite being prefect, he would publish articles in physics of considerable import. He would also be the clear-sighted protector of young Champollion, whose elder brother he knew. See J. Dhombre and J.-B. Robert, *Fourier* (Paris: Belin, 1998).
13. Geoffroy Saint-Hilaire's intervention is often cited: "Hutchinson, inspired by a certain Hamilton, demands that the collections of scholars be handed over to him inso-

far as they are not considered personal property. [. . .] It is at this point that Geoffroy Saint-Hilaire makes a firm resolve and replies to Hamilton, who has come to demand the collections: 'No, no, we will not obey; your army will only be there in two days. Well! From here out the sacrifice will be consummated. You will then dispose of our persons as seems to you fitting.—No, I tell you, it will not be said that such a sacrifice could be accomplished. We ourselves will burn our treasures. It is fame you want. Well! Count on the memory of history: you too will have burnt a library in Alexandria.'" After H. Laurens, *L'expédition d'Egypte* (Paris: Seuil, 1997). The iconography of publications bearing on the Egyptian expedition has recently been republished in *Description de l'Egypte publiée par les ordres de Napoléon Bonaparte* (Paris: Institut d'Orient, 1988).

14. Mondétour is located in the Chevreuse valley, between Orsay and Les Ulis. There is now an industrial zone in that locality.

15. This dangerous journey, in the midst of the Spanish war, shows us again a facet of the strange destiny of Geoffroy Saint-Hilaire. In fact, in the course of the voyage he rescued a Spanish lady who had been slightly injured when her carriage had overturned. He offered his services, lent her his own equipment, and proceeded to go on foot to the nearest village. But a few days later, near Mérida, Geoffroy Saint-Hilaire and his companions were seized by the Spaniards. They were imprisoned, and the crowd demanded that they be put to death. But the Spanish lady he had rescued was the niece of the governor of Estramadure. She pleaded the cause of the Frenchmen with her uncle. He relented, had them set free, returned their carriage—and was then shot for treason some days later.

16. In 1807 Geoffroy Saint-Hilaire published three essays on fishes: E. Geoffroy Saint-Hilaire, "Premier mémoire sur les Poissons; Où l'on compare les pièces osseuses de leurs nageoires pectorales avec les os de l'extrémité antérieure des autres animaux à vertèbres," *Ann. Mus. Hist. Nat.* 9 (1807): 35; E. Geoffroy Saint-Hilaire, "Second mémoire sur les Poissons: Considérations sur l'os furculaire, une des pièces de la nageoire pectorale," *Ann. Mus. Hist. Nat.* 9 (1807): 469; E. Geoffroy Saint-Hilaire, "Troisième mémoire sur les Poissons: Où l'on traite de leur sternum, sous le point de vue de sa détermination et de ses formes générales," *Ann. Mus. Hist. Nat.* 10 (1807): 87.

17. The year 1810 was also when Geoffroy Saint-Hilaire defended his doctoral thesis. Curiously, there is no trace of this event. However, one might ask if this task was not in fact the starting point for a profound meditation whose fulfillment would be the *Anatomical Philosophy* of 1818, his true thesis in every sense of the term.

18. In 1822 Serres was chief physician at the Hospital of Pity. Among other things, he was interested in the comparative anatomy of the vertebrate brain. He would teach anatomy at the Museum in 1839. At first he held the chair that had been created for Portal, and which he rechristened "Anatomy and Natural History of Man." He left it in 1855 to succeed Georges Duvernoy in Comparative Anatomy. His first chair was then occupied by Armand de Quatrefages (1810–1892), who called it the Chair of Anthropology.

Paradoxically, Serres, who was one of the first to have formulated clearly the principle of recapitulation, would be deeply antagonistic to the theory of evolution. He would even be a ferocious opponent of it at the end of his career, believing that, in creating gradually more and more complex organisms, God wished to build a kind of pyramid in order to place man at its summit. We are far here from Geoffroy Saint-Hilaire's battle against the Scale of Beings (see chapter 2).

19. Geoffroy Saint-Hilaire published extensively on crocodiles after his return from Egypt. Thus we find four articles in 1803, three more in 1807 (the year of the three essays on fishes), then two more in 1809. At that time he was most particularly interested in the anatomy of their skulls, which would be important at the time of the affair of the crocodiles of Normandy.

20. I have been unable to find the given names and dates of Laurencet. It seems that he was a physician at Lyon who had come to live in Paris after 1820, and who specialized in cerebral anatomy. As to Meyranx, he was a physician from Montpellier who taught natural science at the Lycée Charlemagne.

21. These articles were immediately translated and published in France: J. W. Goethe, "Principes de philosophie zoologique, discutées en mars 1830, au sein de l'Académie des sciences," *Revue médicale française et étrangère*, December 1830, 1–15; J. W. Goethe, "Dernières pages de Goethe expliquant à l'Allemagne les sujets de philosophie naturelle controversés au sein de l'Académie des sciences de Prais," *Rev. encycl.*, first article, pp. 1–15, second article, pp. 16–31, 1830. They are reprinted in P. Tort, *La querelle des analogues* (Paris: Les Introuvables, Editions d'Aujourd'hui, 1983). The passages translated here are taken from Johannn Peter Eckermann, *Gespräche mit Goethe in den letzten Jahren seines Lebens*, 2 vols. (Basel: Birkhäuser, 1945), vol. 2.

Moreover, in the conversations with Eckermann we find the celebrated exchange of Monday, August 2, 1830, between the German poet and Soret—who had then replaced Eckermann:

> The news of the beginning of the July revolution today arrived in Weimar, and caused a great stir. I went to Goethe in the course of the afternoon. "Well," he called when he saw me, "what do you think of this great event? The volcano has erupted; everything is in flames, and it is no longer a negotiation behind closed doors." "A terrible story," I replied. "But what was to be expected under the known circumstances and with such a ministry, but that it would end with the banishment of the present royal family?" "We do not seem to understand one another, my friend," replied Goethe. "I am not talking about those people; with me it's a question of quite different things! I am speaking of the conflict between Cuvier and Geoffroy Saint-Hilaire, so significant for science, which has finally come to a public outbreak in the Academy." (Eckermann, op. cit., 2:700)

In Goethe's various pronouncements, we constantly find Cuvier's argument on the ambiguity of the words used (plan, composition . . .). We shall return to this problem.

The first part of Goethe's analysis, published in his *Principes de philosophie zoologique*, paints the portraits of the two protagonists:

"At a meeting of the French Academy, on February 22 this year, there occurred an important event that cannot fail to have significant consequences. In this shrine of science, where everything proceeds most respectably in the presence of a large audience, where we encounter the moderation, even the hypocrisy, of well-brought-up people, where only moderate responses are made in the event of differences of opinion, where the doubtful is rather set aside than disputed—here there erupts a quarrel on a scientific point, a quarrel that threatens to become personal, but, looked at closely, comes to mean much more.

"What is revealed here is the ever persistent conflict between two ways of thinking, into which the scientific world has long been divided, and which was constantly creeping along between our neighboring naturalists, but which now takes the stage and erupts with remarkable vehemence.

"Two remarkable men, the permanent secretary of the Academy, Baron Cuvier, and a worthy colleague, Geoffroy Saint-Hilaire, rise to oppose one another; the first known to every one, the second of high repute among naturalists. For thirty years colleagues in one and the same institution, they teach natural history at the Jardin des Plantes, both zealously occupied in this vast field, at first working together, but, through the difference in their views, gradually separated and, instead, moving apart from one another.

"Cuvier works indefatigably as distinction maker, the person who describes exactly what lies before him, and achieves mastery over an immeasurable expanse. Geoffroy Saint-Hilaire, on the other hand, is quietly concerned with the analogies of creatures and their secret relationships. The former proceeds from the particular to a whole, which is indeed presupposed, but is regarded as never to be known. The latter cultivates the whole in his inner sense and lives on in the conviction that the particular can be gradually developed from this starting point. But it is important to notice that much that the latter succeeds in demonstrating clearly and distinctly in experience is taken up gladly by the former. In the same way, the latter by no means disdains whatever particulars are decisively forthcoming from the other side. And so they meet at a number of points, though without admitting any reciprocal influence. For the separator, the maker of distinctions, the one who relies on and proceeds from experience, will not admit a preintuition, a premonition of the particular in the whole. To want to know and to be acquainted with what one does not see with one's eyes, what cannot be tangibly represented, he declares quite plainly to be insolence. Still the other protagonist, holding to certain principles, and relying on an elevated guideline, does not want to concede the validity of that mode of operation.

"After that introductory exposition, no one is likely to object if we repeat what we have just said: here there are two different ways of thinking at work, which are usually to be found in human beings separated and so distributed, that, as is the case elsewhere, they are seldom found joined together in the sciences, and, as they are separated, cannot easily be united [in the same person]. Indeed, it goes so far that, if one part can use something from the other, it nevertheless accepts it with a certain reluctance. If we have the history of science and sufficiently long experience before our eyes, we might fear that human nature could scarcely ever be rescued from this conflict. We carry further what we have been saying.

"The distinction maker applies so much acuity of vision, he needs such unbroken attention, to observe the deviations in form, and finally he also needs the decided talent for naming these differences, so that we cannot really be angry at him if he is proud of his ability, if he treasures this method as the uniquely fundamental and correct one.

"Further, if he now sees the fame that has come to him resting on this basis, he might not easily persuade himself to share these acknowledged advantages with another person who, as it seems, has made the work easier for himself to reach a goal, when the laurels should really be offered only for industry, trouble, and persistence.

"Granted, the one who starts from the idea also believes he is allowed to imagine something—he, who knows how to grasp a leading concept, to which experience is gradually subordinated; he who lives in the sure confidence that what has now been found and already promised in advance will certainly occur again in particular cases. For a man so placed, we also have a certain kind of pride, a certain inner feeling of his advantages, if he does not give up on his side, but can least of all tolerate a certain superciliousness that is often shown him from the opposition, if only in a gentle, moderate way.

"However, what makes the conflict irresolvable may well be the following. Since the distinction maker is concerned with the tangible, and can set out what he accomplishes, demanding no unusual views, and so—though this may seem paradoxical—never lectures—he attracts a larger, even a general, public. In contrast, the other protagonist finds himself more or less as a hermit, who cannot always find himself united even with those who concur with him. This antagonism has often occurred in science and the phenomenon must always reemerge, since, as we have just seen, its elements always develop separately alongside one another, and, whenever they touch, they produce an explosion.

"This usually occurs when individuals of different nations, different generations, or otherwise far apart in circumstances affect one another. In the present case, however, we have the remarkable situation, that two men, equally advanced in age, colleagues for thirty-eight years in the same institution, participating for so many years in different directions in the same field, avoiding one another, tolerating one another, each moving ahead on his own, leading an elegant lifestyle, are at last exposed to, and subjected to, an outbreak, a final public confrontation." J. W. Goethe, *Werke* (Akademieausgabe) (Berlin and Leipzig: Bong, 1928), 36:287–290.

22. This anecdote illustrates the expansion of ideas that characterized the scientific world in France at that time. We are still in the wake of the great scholars of the revolutionary era, and the barriers between the disciplines were not yet impermeable. In fact, Ampère had an early interest in anatomy. He published his lectures a year later. See A. M. Ampère, "Leçons de zoonomie au Collège de France," *Gaz. Méd. Paris* 101 (1832).
23. After the death of Cuvier, Geoffroy Saint-Hilaire's blunders are remarkable. The day after his colleague's death, he wrote to the president of the Academy of Sciences to put up his own name as candidate for the permanent secretaryship that had thus become vacant. Nevertheless, at the same time he launched the idea of a subscription to order a bust of Cuvier to be placed opposite that of Buffon.
24. This title was given to Cuvier by the philosopher and politician Pierre-Paul Royer-Collard (1763–1845).
25. Stendhal, *La Vie d'Henri Brulard,* book 2 (Paris: Le Livre du Divan, 1927), ch. 24, p. 27.
26. Legally speaking, his given names were Jean, Léopold, Nicolas, Frédéric. Georges was a family name given him by his mother in memory of a deceased older brother.
27. To be more precise, Cuvier was brought up speaking French. He learned German at fifteen, when he entered the Caroline University. This was a point much discussed at the close of the nineteenth and the beginning of the twentieth century, at the time of Franco-German antagonism. Von Baer in particular had wanted to demonstrate that the Napoleon of Science was German. See E. L. Troussart, *Cuvier et Geoffroy Saint-Hilaire d'après les naturalistes allemands* (Paris: Mercure de France, 1909).
28. In reality, the concept of the Scale of Beings is an Aristotelian idea, which would be ingeniously revived by Leibniz in his *Monadology* in 1714. It was in fact Aristotle who suggested the idea of ranking animals according to a simple *scala naturae,* according to their "degree of perfection." A gradation was established, starting with "nutritive souls," corresponding to plants, to "rational souls," characteristic of man, and perhaps to other sorts, superior to him. In this scale every individual possesses all the powers of those situated below it, augmented by those peculiar to it. This concept survived through the Middle Ages until the eighteenth century and brought about the acceptance of the idea of a great chain of beings, composed of an infinity of links hierarchically arranged, starting with the simplest existences, and leading through all the intermediates to the *ens perfectissimum.* This image impregnates the whole of the eighteenth century, and even appears in poetry, as in these verses of Ecouchard-le-Brun:

    All bodies are linked in the chain of Being.
    Nature everywhere both precedes and follows.
    [. . .]
    Her steps developed in a constant order,
    Never escaping a precipitous border,

From man to animal lessening the distance,
See the Wild Man link their existence.
From the dubious coral, neither plant nor mineral,
Return to the Polyp, an insect that is vegetal.

This eighteenth-century concept was totally destroyed by Darwinism. Unfortunately, the idea remains anchored in some minds. For a deeper analysis see A. O. Lovejoy, *The Great Chain of Being*, 13th ed. (Cambridge: Harvard University Press, 1976; originally published 1936); P. Mengal, *Histoire du concept de récapitulation: Ontogenèse et phylogenèse en biologie et sciences humaines* (Paris: Masson, 1993).

29. Cuvier succeeded Professor Parrot, a native of Montbéliard and a friend of his parents. Thus at Caen he became tutor to the son of the count of Héricy, Achille, who was born July 10, 1776. It appears that Achille was a very mediocre student, preferring rides around the countryside to Latin essays.

30. Achille's grandfather, the marquis of Héricy, was governor of a part of Lower Normandy. Very wealthy, he kept a superb botanic garden at Caen. Starting in 1792, when the political scene became troubling, the Héricy family no longer left the chateau of Fiquainville, in the district of Caux, near the communes of Bec aux Cauchois. In Cuvier's zoological diary (*Diarum zoologicum*), we find traces of the study of seventeen mollusks, all now very classic (mussel, oyster, cockle, clam, razor clam, limpet . . .). The key point is that, unlike his contemporaries, he is not satisfied to observe their shells. He is interested in the soft parts, which constitute the real body of the animal. He also studied crustaceans. There are fifteen to be counted, equally classic: various crabs (spider crab, velvet-backed swimming crab, edible crab, velvet fiddler . . .), shrimps, a lobster, a hermit crab.

Further information can be found in G. F. Dollfus, "Le séjour de G. Cuvier en Normandie," *Bull. Soc. Linn. Norm. Ann. 1925*, 1926, 156–178.

31. See Cuvier's letter to Pfaff, November 17, 1788, in G. L. Cuvier, *Lettres à C. M. Pfaff sur l'histoire naturelle, la politique et la littérature, 1788–1792*, trans. L. Marchaant (Paris: Librairie Victor Masson, 1858), 73.

32. The former Abbé Tessier, of the Academy of Sciences, and of the Society of Medicine and Agriculture, was then directing a hospital at Fécamp in which he had a school of military health. He had asked Cuvier to give lectures there in botany.

Alexandre Tessier had the title of "abbé" because he had received a scholarship from the archbishop, and was in lower orders, but he had never been ordained a priest. He is known for his study of ergot in rye and for the introduction of merino sheep. The present Lycée Jean-Baptiste Say, in the sixteenth arrondissement of Paris, is the (remodeled) former sheep fold in which Daubenton raised sheep for the Abbé Tessier.

33. This would be the theme of the *Tableau élémentaire de l'histoire naturelle des animaux,* which Cuvier was to publish in 1798.

34. At that period, the Central Schools played the role of lycées for a short time. The

Central School of the Pantheon then occupied the quarters of the present Lycée Henri IV.

35. The iconography published for the centenary of Cuvier's death is fascinating. There we see Cuvier, delicate and suffering in his youth, becoming more and more imposing with age and honors. See L. Bultingaire, "Iconographie de Georges Cuvier," *Archives du Muséum*, 6th ser., 9 (1932): 1–12.

36. On July 29, Monseigneur de Quélen had taken refuge at the Hospital of Pity, in Serres's section. He was found there by a group of hostile demonstrators. Serres asked Geoffroy Saint-Hilaire to hide the prelate, even though he was favorable to the July Revolution. On the July 31, the archbishop, in disguise, reached Geoffroy Saint-Hilaire's residence, and stayed there until August 14.

37. This would be undeniably one of the deep causes of Cuvier's fixism, but also one of the motives for Geoffroy Saint-Hilaire's transformism, which, in the spirit of the time, incorporated the ideas of "progress" put forward by the Saint-Simonians and Fourierists.

38. Opinions are divided on the question whether Cuvier in fact died of cholera, or of an acute myelitis during the cholera epidemic. Louis Roule leans toward the first hypothesis. See L. Roule, "La vie, la carrière et la mort de Cuvier," *Archives du Muséum*, 6th ser., 9 (1932): 13–20.

39. R. Anthony, "Cuvier et la Chaire d'Anatomie Comparée du Muséum National d'Histoire Naturelle," *Archives du Muséum*, 6th ser., 9 (1932): 21–31.

40. Introduction to the *Recherches sur les ossemens fossiles*, translated in M. J. S. Rudwick, *Georges Cuvier, Fossil Bones, and Geological Catastrophes* (Chicago: University of Chicago Press, 1997), 217.

41. Georges Cuvier, *Le Règne animal* (Paris: Deterville, 1817), 1:10.

42. Rudwick, loc. cit.

43. "Discour prononcé par le citoyen Cuvier à l'ouverture du cours d'anatomie comparée qu'il fait au Muséum de l'histoire naturelle."

44. Cuvier, *Régne animal*, 1:57.

45. Rudwick, loc. cit., 220.

46. Honoré de Balzac, *La Comédie humaine*, vol. 9, *Etudes philosophiques* I, *Le Peau du chagrin* (Paris: La Pléiade, 1937), p. 29.

47. Pietro Perugino (ca. 1448–1521) was the master of Raphael (1483–1520), who is thought to have brought to completion the work initiated by the older painter.

48. Rudwick, loc. cit., 190.

49. Ibid., 228–229.

50. Ibid., 229.

51. Notes were taken by Magdeleine de Saint-Agy. See M. L. Roule, "Cuvier historien scientifique," *Archives du Muséum*, 6th ser., 9 (1932): 77–82.

52. According to Patrick Tort: "There is no doubt that Geoffroy Saint-Hilaire's choice of such a title as 'Principles of Zoological Philosophy' to crown the sum total of the speeches made in the course of the academic debates, and especially the results of his

own research in the light of the theory of analogues, had a deep relation to the priority of the *Zoological Philosophy* of Lamarck (1809)." C. P. Tort, *La Querelle des analogues* (Paris: Les Introuvables, Editions d'Aujourd'hui, 1983), 18.

53. *Le Journal des débats* leaned toward Cuvier. On the other hand, *Le Temps* and *Le National* defended Geoffroy Saint-Hilaire. Moreover, the *Principes de philosophie zoologique* could not have been published so quickly if Geoffroy Saint-Hilaire had not reprinted in it several of his articles. One of the editors of *Le National* was none other than Adolphe Thiers.

54. Ernest Renan (1823–1892) published this work at the end of his life, in 1890, but it was a youthful text, written in 1848–1849. The exact title is *L'Avenir de la science, pensées de 1848 (The future of science: thoughts of 1848).*

Chapter Two

1. The complete text of these two works was reproduced in 1968 by Culture et Civilisation in Brussels.

2. We may remark that Etienne is the only member of the family to have added "Saint-Hilaire" to his name. According to Théophile Cahn, this surname was given him by his family when he was young. Marc Antoine received the surname Château. The descendants of the two brothers would have the names of Geoffroy Saint-Hilaire and Geoffroy Château respectively. Cuvier, like Geoffroy, experienced adventures with his name. The fixist kept, as a mark of continuity, the given name of a deceased elder brother (see chapter 1, n. 26), and the transformist belonged to a family in the bosom of which younger generations distinguished themselves by new names.

3. Etienne returned from Egypt with his brother on the ship reserved for engineering officers.

4. The dedicatory page carries this footnote: "Assist, love, and adopt my young liberator," words of Haüy to Daubenton, after he had evaded the decrees of September 2 and 3. The celebrated Daubenton adopted and became an associate of his friend's young liberator.

5. Given his own break with Buffon, Daubenton knew what Geoffroy Saint-Hilaire was exposing himself to. Goethe commented subtly on the parallel between Buffon and Geoffroy Saint-Hilaire on the one hand and Daubenton and Cuvier on the other:

> Buffon takes the external world in its diversity as he finds it, as a coherent, persistent whole, interacting in reciprocal respects. Daubenton, as an anatomist, constantly occupied in separating and sorting, hesitates to unify anything he finds on its own with something else. He sets each thing carefully alongside another, and describes each one for itself.
>
> Cuvier works in the same way, but with more freedom and more vigilance. To him is given the gift of observing unlimited particularities,

of distinguishing, of comparing, of placing and ordering them, and so of earning great honor.

But he too has a certain grasp of a higher method, which after all he cannot do without, and which he applies, though unconsciously. And so he repeats Daubenton's characteristics at a higher level. Similarly, we might say that Geoffroy to some extent points back to Buffon. For if Buffon acknowledges and assimilates the great synthesis of the empirical world, but at the same time pays attention to and makes use of all the criteria that present themselves to him on behalf of differentiation, Geoffroy approaches more closely to the great abstract unity of which Buffon had had only a presentiment. It does not frighten him, and when he grasps it, he knows how to use its consequences to his advantage. (J. W. Goethe, *Principes de philosophie zoologique* in *Werke* [Berlin and Leipzig: Bong, 1928], 36:298)

6. E. Geoffroy Saint-Hilaire, *Mémoire sur les rapports naturels des Makis Lémur L.*
7. Darwin set sail December 27, 1831, and returned October 2, 1836. An account of this long journey is given in *The Voyage of the Beagle* (Garden City, N.Y.: Doubleday, 1962, based on the 1860 edition). See also *The Autobiography of Charles Darwin*, ed. N. Barlow (London: Collins, 1958).
8. E. Geoffroy Saint-Hilaire, "Considérations sur les pièces de la tête osseuse des animaux vertébrés, et particulièrement sur celles du crâne des Oiseaux," *Ann. Mus. Hist. Nat.* 10 (1807): 342. The spirit of the method would be restated in integral fashion in the *Philosophie anatomique* of 1818.
9. G. Cuvier, "Sur la composition de la tête osseuse dans les animaux vertébrés," *Nouv. Bull. Sc. Sté. Phil.* 3 (1812): 117.
10. Thomas Kuhn, *The Structure of Scientific Revolutions* (Chicago: University of Chicago Press, 1962).
11. As we have already noted, the first volume of the *Histoire des poissons* of Cuvier and Valenciennes would be published only in 1828. See L. Roule, "Cuvier ichtyologiste," *Archives du Muséum*, 6th ser., 9 (1932): 47–54; G. Lecointre, "Aspects historiques et heuristiques de l'ichtyologie systématique," *Cybium* 18 (4) (1994): 339–430.
12. The Batrachia, which are also tetrapods, were only seldom studied by Geoffroy Saint-Hilaire.
13. See chapter 1, n. 28.
14. Etienne Serres must certainly have been a major influence. Moreover, he is thanked in the preface "for having communicated his precious manuscripts on the encephalon and on the formation of the bony tissues." It is probably a question of the manuscript of the *Recherches physiologiques et pathologiques sur le cervelet de l'homme et des animaux*, published in 1823.
15. The original texts are: E. Geoffroy Saint-Hilaire, "Mémoire sur l'organisation des insectes: Premier mémoire sur un squelette chez les Insectes dont toutes les pièces

identiques entre elles dans les divers ordres du système entomologique correspondent à chacun des os du squelette dans les classes supérieures (lu à l'Académie des Sciences, le 3 janvier 1820)," *J. compl. du dict. des sc. méd.* 5 (1820): 340; E. Geoffroy Saint-Hilaire, "Mémoire sur l'organisation des insectes: Second mémoire, sur quelques règles fondamentales en philosophie naturelle (lu à l'Académie des Sciences le 17 janvier 1820)," *J. compl. des sc. méd.* 6 (1820): 30; E. Geoffroy Saint-Hilaire, "Mémoire sur l'organisation des insectes: troisième mémoire, sur une colonne vertébrale et ses côtes dans les insectes apiropodes (lu à l'Académie des Sciences le 12 février 1820)," *J. compl. des sc. méd.* 7 (1820): 271.

The republished edition, which is better known, is E. Geoffroy Saint-Hilaire, *Mémoires sur l'organisation des insectes* (Paris: Crevot, 1814). Finally, Geoffroy Saint-Hilaire restated his ideas in the memoir of 1822, which has recently been much cited: E. Geoffroy Saint-Hilaire, "Considérations générales sur la vértèbres," *Mém. Mus. Hist. Nat.* 9 (1822): 89–114.

16. The concept of *composition* is of central importance in the work of Geoffroy Saint-Hilaire. It has been constantly criticized, even by experts who were on his side. This is how Goethe, in his conversation with Eckermann on June 20, 1831, tried to explain the controversy between Cuvier and Geoffroy in terms of a problem with a word: "Geoffroy Saint-Hilaire is a man who really has great insight into the spiritual reign and working of nature; but his French language, insofar as he is forced to make use of its idiomatic expressions, keeps leaving him in the lurch—and this not only for obscurely contrived, but also with wholly visible, purely physical objects and relations. . . . Just as inappropriately . . . , the French use the expression *composition* when they talk about the events of nature. True, I can indeed put together the separate parts of a machine that is made piece by piece, and I can speak of composition in the case of such an object, but not if I have in mind the individual parts of an organic whole, which form themselves as living things and are penetrated by a single soul. . . . How can we say that Mozart *composed* his *Don Juan*? *Composition!* As if it were a piece of cake or a cookie, that is stirred up out of eggs, flour, and sugar!" J. P. Eckermann, *Gespräche mit Goethe* (Basel: Birkhäuser, 1945), 2:709.

However, a deeper study of the use of the word "composition" does not produce that resonance. The fact is that one sense of the term goes back to the work of art, as introduced chiefly by Alberti in his *De Pictura* of 1645: "Composition is the procedure in painting whereby the parts are composed together in the picture. The great work of the painter is not a colossus but a 'historia,' for there is far more merit in a 'historia' than in a colossus. Parts of the 'historia' are the bodies, part of the body is a member, and part of the member is the surface. The principal parts of the works are the surfaces, because from these come the member, from the members the bodies, from the bodies the 'historia,' and finally the finished work of the painter." Leon Battista Alberti, *On Painting and On Sculpture*, ed. and trans. Cecil Grayson (London: Phaidon, 1972), 73.

Without the right approach to reading it, this is a somewhat enigmatic statement. But let us notice at once that we find in this text on the one hand the notion of levels of organization—in biology, cells form tissues; tissues form organs; organs form organisms—and on the other hand the notion of unity, history—in the original sense of the term, that isscenography, for Alberti, the harmony of the organism in biology.

The reading given by Michael Baxandall is particularly illuminating. He first restates the definition of composition as "the way in which a painting can be organized so that each plane surface and each object plays its part in the effect of the whole." Michael Baxandall, *Giotto and the Orators: Humanist Observers of Painting in Italy and the Discovery of Pictorial Composition, 1350–1450* (Oxford: Oxford University Press, 1971), 129. Let us notice at once that, in order to explain what composition is, the author uses the verb "organize." But this verb carries in itself the idea of someone who does the organizing, and of a goal for the organization. On the contrary, the term "composition" has much less the meaning of finality, and thus seems to be able to be used in biology.

Baxandall goes on to ask why Alberti felt it necessary to explain the concept of composition. "Alberti urged high standards of relevance and organization in a climate of public taste, particularly humanist taste, that often favoured painting less rigorous in this respect" (Baxandall, 130). Here again we find a parallel with Geoffroy Saint-Hilaire, who was struggling against the facility of certain anatomists, as we can see in the two preliminary discourses.

Finally, Baxandall explains clearly that Alberti had not invented these levels of organization: "for the notion of *compositio* is a very precise metaphor transferring to painting a model of organization derived from rhetoric itself. *Compositio* was a technical concept every schoolboy in a humanist school had been taught to apply to language. It did not mean what we mean by literary composition, but rather the putting together of the single evolved sentence or period, this being done within the framework of a four-level hierarchy of elements: words go to make up phrases, phrases to make clauses, clauses to make sentences" (Baxandall, 131). Alberti simply transposed rhetoric to painting; the words are the plane surfaces; the phrases, the members; the clauses, the bodies; the sentence, the painting. We can say that Geoffroy Saint-Hilaire, whether consciously or not, transposed painting to biology. Then the concept of *organic composition* assumes all its dimensions.

Chapter Three

1. E. Geoffroy Saint-Hilaire, *Mémoires sur l'organisation des insectes* (Paris: Crevot, 1824).
2. See the references in chapter 2, n. 15.
3. Rathke is famous in embryology for having demonstrated that a mammalian embryo presented, at one moment of its development, branchial pouches homologous to the branchial slits of fishing, testifying to the fact that a distant ancestor must have had

an aquatic life. See L. K. Nyhart, *Biology Takes Form: Animal Morphology and the German Universities (1800–1900)* (Chicago: University of Chicago Press, 1995).

4. E. Geoffroy Saint-Hilaire, *Notions synthétiques, historiques et physiologiques de philosophie naturelle* (Paris: Denan, 1838).
5. E. Geoffroy Saint-Hilaire, "Considérations générales sur la vertèbre," *Mém. Mus. Hist. Nat.* 9 (1822): 89–114. This note is the sequel to two articles that precede it in volume 9 of the *Mémoire du Museum:* "Sur une nouvelle espèce de Boeuf, nommé Gaour par les Indiens, d'une taille gigantesque et ayant les apophyses épineuses des vertèbres dorsales prolongées extérieurement," Mém. Mus. Hist. Nat. 9 (1822): 71–75; "Sur les tiges montantes des vertèbres dorsales, pièces restreintes dans les mammifères à un état rudimentaire et portées chez les Poissons au maximum du développement, pour servir à l'intelligence de la notice sur le Gaour," *Mém. Mus. Hist. Nat.* 9 (1822): 76–78.
6. The central figure of plate 6 was reproduced by S. J. Gould in his article in *Natural History* in 1985, and figure 2 of plate 7 by de Robertis and Sasai in their article in *Nature*, vol. 380 (1966).

Chapter Four

1. G. Cuvier, *Recherches sur les ossemens fossiles* (Paris: Dufour et Ocagne, 1825).
2. During the war of 1914–1918 the name of this village was changed by a decree of the municipal council. It is now called Fleury-sur-Orne.
3. Lamarck published his *Philosophie zoologique* in 1809. Remember what Cuvier thought of Lamarck's system: "a system resting on such foundations may amuse the imagination of a poet: a metaphysician can derive from it a whole further generation of systems; but it can bear the examination of any one who has dissected." Cited in J. Piveteau, "Le débat entre Cuvier et Geoffroy Saint-Hilaire sur l'unité du plan et de composition," *Rev. Hist. Sci.* 3 (1950): 342–368.
4. E. Geoffroy Saint-Hilaire, "Recherches sur l'organisation des gavials: Sur leurs affinités naturelles, desquelles résulte la nécessité d'une autre distribution générique, *Gavialis, Teleosaurus* et *Steneosaurus;* et sur cette question, si les Gavials *(Gavialis)*, aujourd'hui répandus dans les parties orientales de l'Asie, descendent, par voie non interrompue de génération, des Gavials antédiluviens, soit des Gavials fossiles, dits Crocodiles de Caen (*Teleosaurus*), soit des Gavials fossiles du Havre et de Honfleur (*Steneosaurus*)," *Mém. Mus. Hist. Nat.*, 3 (1825): 97, and *Nouv. Bull. Philom.* 13 (1825).
5. Geoffroy Saint-Hilaire's note is here reproduced in full. (Don't forget that in 1825, Lamarck was eighty-one years old and would die four years later.)

> We shall invoke the spirit of those two laws (see *Philosophie zoologique*, 1:235, Paris: published by the author at the Royal Garden, 1809).
>
> *First law:* In every animal that has not passed beyond the term of its development, the more frequent or the different or the earlier use of any

organ strengthens that organ, develops it, enlarges it, and gives it a power proportionate to the interest of that action, etc.

> *Second law:* All that nature has made individuals acquire or lose through the influence of the circumstances to which their race has found itself exposed is perpetuated through the path of generation, etc.
>
> One cannot recommend too highly for the meditation of young people the perusal of the philosophical exposition (seventeen pages) that precedes these conclusions. The author has taken his views through an order of facts and necessary consequences, but not with respect to the applications that immediately follow. All the variations described after p. 235, followed by explanations, seem to me to depend instead on very different primitive facts, to wit: on alterations consequent on the distribution of the arteries; alterations that in their turn depend on other causes, some of which I believe I have specified in my work on human monstrosities.
>
> I see certain explanations flowing from closely linked and necessary facts, in the same way as Newton, when he attributed to choice the marvelous uniformity of the planetary system, and added, as a development of his thought at the same time both philosophical and religious, that "every artifice of a like uniformity, like the uniformity of animal composition, could only be the work of the wisdom and intelligence of a powerful Agent, who, through the fact that he is present everywhere, is all the more capable of moving the body by his will in his uniform and infinite *Sensorium*" *(Opticks).*

However, despite this optimistic note, Geoffroy Saint-Hilaire would later cite Lamarck only very episodically. It is true that their ideas are very different. As Bourdier emphasizes, "In conformity with classical materialism and contrary to Lamarck, Geoffroy Saint-Hilaire believes that it is the organs that, by their structure, determine the functions." See F. Bourdier, "Lamarck et Geoffroy Saint-Hilaire face au problème de l'évolution biologique," *Revue d'histoire des sciences,* 1972, pp. 311–325.

6. There exists in the archives of the Academy of Sciences an eloquent manuscript of Geoffroy Saint-Hilaire's, which is a sketch for a reply to the article "Nature" published in 1825 by Cuvier in Levrault's *Dictionnaire d'histoire naturelle:*

> It is indeed deplorable to have an engagement of this nature with one of the leading scholars of this period, with an old comrade, whom my active solicitude for the sciences at the beginning of my career led me to have discovered and sought out in the depths of our provinces, with a colleague, finally, with regard to whom I have been making a public profession of friendship for twenty-five years. Baron Cuvier seems to be struck

> by a general enthusiasm that he considers fatal to the true march of the sciences: in fact, considerable works, including those that are becoming the manual of medical youth (note by G.S.-H.: *Manuel d'anatomie* by F. Meckel, French translation, vol. 1, p. 2). Will M. Cuvier's opposition stop this enthusiasm of minds? . . . In 1810, I agree, the identity of organic composition could pass for a hypothetical idea that was more or less probable and perhaps it would then already be considered one of the happier speculations of the mind. But in 1825 it is a general fact, like attraction; it is a fact of the same order, of equal importance, which has entered in the same way into the domain of the mind.

See Y. Laissus, "Catalogue des manuscrits d'Etienne Geoffroy Saint-Hilaire conservés aux Archives de l'Académie des Sciences de l'Institut de France," *Histoire et nature* 1 (1973): 71–88.

7. J. Rostand, "Etienne Geoffroy Saint-Hilaire et la tératogenèse," Centre International de Synthèse, *Revue d'histoire des sciences* 17, no. 1 (1966): 41–50.
8. Bourdier quotes this statement of Cuvier's in his explanation of the Baron's distress:

> Cuvier in his *Histoire des progrès des sciences naturelles* (ed. 1873, 2:327–28) informs us that in the following year Geoffroy tried to verify the views of the late Lamarck, and a certain antediluvian geology, by undertaking experiments on the hens' egg in the course of its development, trying to draw its organization into unaccustomed paths; he adds, "Geology in particular, if it succeeds in modifying a single species, will itself be radically modified in one of its principal foundations." Cuvier, it seems, was then very much afraid that Geoffroy might succeed and he would have tried to have his research forbidden by the government.

See F. Bourdier, "Quelques aspects sur la paléontologie évolutive en France avant Darwin," *Bull. Soc. Géol. France* 1 (1959): 881–896.

9. "I had thought that some experiments in physiology might be undertaken for the benefit of antediluvian geology. . . . This was the only means of knowing if organs are modified, and if, in changing some into others, they have not, by this very fact, undergone an infinite sequence of diversities. But I have come to believe that experiment, carried out on a grand scale, would not give the desired results." Let us not forget that we are situated here before the discovery of genetics. Geoffroy Saint-Hilaire did not produce a mutation by his experiments. The variations he obtained were not hereditary. But they are interesting to the extent that they exhibit the *field of possibles,* starting from a given plan of organization. See E. Geoffroy-Saint-Hilaire, "Mémoire où l'on se propose de rechercher dans quels rapports de structure organique et de parenté sont entre eux les animaux des âges historiques et vivant actuellement, et les espèces antédiluviennes et perdues," *Mémoires du Muséum* 17 (1829): 200.

10. For today's science, the affair is closed. Both forms are in the family of the *Teleosauridae* within the *Crocodilia*. The forms initially described are now:

- *Steneosaurus* Geoffroy 1825: *Steneosaurus rostro-minor* Geoffroy, *Metriorhynchus geoffroyi; Steneosaurus rostro-major* Geoffroy; *Steneosaurus magistrorhynchus* (names altered by Deslongchamps in 1866)
- *Teleosaurus* Geoffroy 1825: *Teleosaurus cadomensis* (Lamouroux, 1820) (originally described under the name *crocodilus*).

The difference between these animals is minimal, since both belong to the same family. Thus Geoffroy Saint-Hilaire's interpretations were quite exaggerated.

Chapter Five

1. Louis Lambert, a well-known hero of Balzac's, encountered Meyranx "at the course in comparative anatomy and in the galleries of the Museum, both led there by a single subject of study, the unity of zoological composition."
2. The text was never published. It was summarized and commented on by Cuvier after his intervention of February 22, which would be published almost immediately: G. Cuvier, "Quelques considérations sur l'organisation des Mollusques," *Annales des Sciences Naturelles* 19 (1830): 241–259. Geoffroy Saint-Hilaire gave what was clearly a summary of it in his *Philosophie zoologique.*
3. *Plicature:* the term indicates pictorially how the two biologists saw the organization of the cuttlefish. They compared the animal to an acrobat who *folds* himself up at the level of the pelvis, in such a way that his legs come close to his arms.
4. Paradoxically, Cuvier here echoes an argument of Goethe's, although the latter was a partisan of Geoffroy Saint-Hilaire (see chapter 2, n. 16). In the three articles presenting the controversy of 1830 to the German public, and translated into French by Geoffroy Saint-Hilaire, he drew attention to the use of a natural language, in this case, French. Goethe showed that the principal terms used by the two protagonists were for the most part equivocal. Speaking of Geoffroy Saint-Hilaire, he said: "Unfortunately, on many points his language does not offer him a precise expression. Since his adversary finds himself in the same situation, the controversy becomes unclear and confused." He then analyzes successively the meaning of "matériaux," "composition," "embranchement," and "plan" and demonstrates their misuse. Of "composition" he remarks: "This is again an unhappy word." Discussing "embranchement," he remarks

> We believe we see here, in the particular case as in the whole, the influence of that epoch in which the nation (i.e., the French nation) was devoted to sensualism, accustomed to use material, mechanical, atomistic expressions; for if the inherited linguistic usage is indeed adequate for ordinary

conversation, as soon as the discourse rises to the intellectual level, it clearly contradicts the higher views of superior men. (J. W. Goethe, *Werke* [Akademieausgabe] [Berlin and Leipzig: Bong, 1928], 36:309–10)

5. See the article "Philosophie de la Nature" in P. Tort, ed., *Dictionnaire du Darwinisme et de l'évolution* (Paris: Presses Universitaires de France, 1996), 3418–3425.
6. Or rather "potentially," if we adopt the terminology defended by Stéphane Lupasco. Cf. H. Guyader, *Théorie et histoire en biologie* (Vrin: Paris, 1988); S. Lupasco, *Le Principe d'antagonisme et la logique de l'énergie* (Paris: Hermann, 1951).
7. E. Serres, *Précis d'anatomie transcendante appliquée à la physiologie* (Paris: Librairie Charles Gosselin, 1842), vol. 1.
8. Ibid.
9. The concept of this law of Geoffroy Saint-Hilaire-Serres-Meckel is difficult to present in integrated fashion, and it is especially important not to analyze it according to present-day concepts. Thus Daudin comments:

> The idea of Geoffroy and Serres is not that the individual is constrained to reproduce by epitomizing the history of its ancestors in virtue of physical dispositions that it acquires from that source, but rather that one and the same course of development comprising the whole hierarchy of possible forms lies open for the genesis of animals and that the only difference between them will consist in their stopping at a later or earlier point. Instead of the past of the race, transmitted by heredity, tracing in the embryo the paths of its differentiation, it is the necessary series of stages of progression, irrecusably marked in the order of embryonic developments, that imposes the result of these developments, by the same authority, on all zoological forms.

Ernst Haeckel's law of 1866, following the Darwinism of the period, will be entirely historical:

> The series of forms by which the individual organism passes from the primordial cell to its full development is only a repetition in miniature of the long series of transformations undergone by the ancestors of the same organism from the most remote times to our own day.

We now summarize it "ontogeny recapitulates phylogeny."

The difference between the two laws is essential: the first law considers neither the transmission of characters from generation to generation nor the causes of variations. The second law introduces the time scale of phylogeny and takes account only of hereditary characters. Considering only the past, it does not presuppose the notion of finalism.

See G. Canguilhem et al., "Du développement à l'évolution au XIXe siècle," *Thales* 11 (1960); H. Daudin, *Les Classes zoologiques et l'idée de série animale* (1926; reprint, Paris: Presses Universitaires de la France, 1983); S. J. Gould, *Ontogeny and Phylogeny* (Cambridge: Harvard University Press, 1977); P. Mengal, *Historie du concept de récapitulation: Ontogenèse et phylogenèse en biologie et sciences humaines* (Paris: Masson, 1993).

10. H. Le Guyader, *Théorie et histoire en biologie* (Paris: Vrin, 1988).
11. At the close of the eighteenth century, the philosophy of nature of Friedrich Schelling (1775–1854) had imagined that animals, from the most complex to the simplest, were potential men halted in their development at earlier and earlier stages. Cf. F. Bourdier, "Trois siècles d'hypothèses sur l'origine et la transformation des êtres vivants (1550–1859)," *Revue d'histoire des sciences* 13 (1960): 1–43.

Chapter Six

1. Charles X mounted the throne in September 1824, on the death of Louis XVIII. In 1827, the new chamber is hostile to the Villèle government. In January 1828, Charles X replaces Villèle with Martignac, a royalist too moderate for his taste. He removes him quickly, and on August 8, 1829, forms a new government "after his own heart" under the direction of the prince of Polignac, with incompetent and unpopular ministers in the key posts (Polignac in Foreign Affairs, La Bourdonnaye in the Interior, Bourmont in War, Monbel in Public Instruction, Chabrol de Crouzol in Finance). In March 1830—amid full controversy—the Chamber of Deputies recalls in an address the inalienable rights of national representation. It is dissolved on May 16, and the government sets new elections on June 23 and July 3, from which the liberal opposition emerges in greater strength. Charles X responds on July 25 with a show of force in the ordinances of Saint-Cloud, suspending the freedom of the press, dissolving the newly elected Chamber, modifying the electoral law, and postponing new elections to September. Paris rises immediately. On the third day of the Three Glorious Days—July 27, 28, and 29, 1830—the insurgents seize the advantage. The Louvre falls and Marmont's troops take flight. Watching them run from his window in the rue Saint Florentin, Talleyrand, according to tradition, is said to have consulted his watch and said: "*At five minutes past twelve, the elder branch of the Bourbons stopped reigning.*" On the morning of July 30, the troops leave Paris. A proclamation composed by Thiers (then editor in chief of the liberal paper *Le National*) proposes naming as king the duke of Orleans, son of Phillippe-Egalité. Lafayette, at the time commander of the National Guard, joins in and on July 31 presents the duke of Orleans on the balcony of the Hôtel de Ville. On August 3, Charles X, who has taken refuge at Rambouillet, names the duke of Orleans lieutenant general of the kingdom, and abdicates in favor of his grandson, the duke of Bordeaux. But the liberal deputies declare the throne vacant, thus thrusting aside Henry V. The Charter is reviewed and approved on August 7. The duke of Orleans is proclaimed Louis-

Philippe I, king of the French. He is sworn in on August 9. See J. Tulard, *Histoire de France: Les révolutions, de 1789 à 1851* (Paris: Fayard, 1985).

2. In 1831 Geoffroy Saint-Hilaire published a work bringing together the five memoirs that had as their theme extant and fossil crocodiles: E. Geoffroy Saint-Hilaire, *Recherches sur de grands sauriens trouvés à l'état fossile vers les confins maritimes de la Basse-Normandie, attribués d'abord au Crocodile, puis déterminés sous les noms de Téléosaurus et Sténéosaurus* (*Investigations on the great saurians found in the fossil state near the maritime borders of Lower Normandy, first attributed to the crocodile, but then identified under the names of Teleosaurus and Steneosaurus*) (Paris: Firmin-Didot, 1831).

   This collection includes:

   First Memoir (read to the Royal Academy of Sciences on October 4, 1830): "Sur les lames osseuses du palais dans les principales familles d'animaux vertébrés, et en particulier sur la spécialité de leur forme chez les crocodiles et les reptiles téléosauriens" ("On the bony plates of the palate in the principal families of vertebrates, and in particular on their special form in the crocodiles and teleosaurian reptiles").

   Second Memoir (read October 11, 1830): "La spécialité des formes de l'arrière-crâne chez les Crocodiles, et l'identité des mêmes parties organiques chez les reptiles téléosauriens" ("The special forms of the hind-skull in the crocodiles and the identity of the same organic parts in the teleosaurian reptiles").

   Third Memoir (read May 9, 1831): "Des recherches faites dans les carrières du calcaire oolithique de Caen, ayant donné lieu à la découverte de plusieurs beaux échantillons et de nouvelles espèces de Téléosaurus" ("On investigations made in the quarries of the oolithic limestone of Caen, which gave rise to the discovery of several beautiful specimens of new species of Teleosaurus").

   Fourth Memoir (read March 28, 1831): "Le degré d'influence du monde ambiant pour modifier les formes animales; question intéressant l'origine des espèces téléosauriennes et successivement celle des animaux de l'époque actuelle" ("The degree of influence of the environment in modifying animal forms; a question having to do with the origin of teleosaurian species and successively that of the animals of the present epoch").

   Fifth Memoir (read August 29, 1831): "Les pièces osseuses de l'oreille chez les crocodiles et les reptiles téléosauriens retrouvés en même nombre et remplissant les mêmes fonctions que chez tous les autres animaux vertébrés" ("The bony pieces of the ear in the crocodiles and teleosaurian reptiles recovered in the same number and fulfilling the same functions as in all the other vertebrates").

3. In this fourth memoir, Geoffroy Saint-Hilaire uses two terms that will later be important: *mutation* and *evolution*, but clearly in a sense different from that of genetics: "*What we do not yet understand, and consequently what we have at present to search for, is how, under the power of contemporary physical conditions and of analogous facts, the mutation of organization is really possible, how it was and has to have been formerly prac-*

ticable. . . . Let us look further, and try to understand the attraction of the partisans of the system of evolution."

It is traditional to attribute to Spencer, in 1852, the first use of the word *evolution* in its modern meaning. But Frédéric Gérard frequently uses this term as early as 1844. We can push further back in time and note that Louis François Charles Girou de Buzareingues (1773–1856) already uses it as early as 1828 in his book *De la Génération*. Geoffroy Saint-Hilaire uses the term here, and then in an article in 1833, thus well before Spencer—with a modern vision: *"Of the two theories on the development of organs, one supposes the preexistence of the germs and their indefinite nesting; the other admits their successive formation and their evolution in the course of ages."* Cf. F. Bourdier, "Trois siècles d'hypothèses sur l'origine et la transformation des êtres vivants (1550–1859)," *Revue d'histoire des sciences* 13 (1960): 1–44; E. Geoffroy Saint-Hilaire, "Considérations sur des ossemens fossiles, la plupart inconnus, trouvés observés dans les bassins de l'Auvergne," *Rev. encycl.* 59 (1833): 315.

4. Geoffroy Saint-Hilaire's research in this domain extends over six years, from 1820 to 1826. He uses an artificial incubating oven first in Auteuil, then at Bourg-la-Reine. Following his ideas on the modifying action of the respiratory medium, he tries to make it vary artificially. In fact, Geoffroy Saint-Hilaire is trying to reconstitute the natural conditions of monstrosity, as he has said in an unpublished manuscript at the Museum: "*In order to try to meet one of the circumstances proper to ordinary incubations, I had to keep eggs accessible to the heat of the mother in all the variable conditions, and perhaps in this way we would encounter the causes of true monstrosities.*"

Camille Dareste (1822–1899), who later takes up again Geoffroy Saint-Hilaire's works in teratology, is very explicit. In 1877, he will write: "Would it not be possible to obtain, by the use of methods analogous to those I have employed, simple varieties of organization compatible with life and reproduction, and consequently capable of becoming hereditary, and consequently of constituting the point of departure for true races? Thus we would succeed through experimentation in raising the greatest problem, not only of zoology, but even of natural history as a whole, that of the origin of the forms of life, of knowing if they are absolutely fixed or indefinitely variable. This was the thought that was guiding me when I undertook that long series of teratological experiments, full of promise for the future, to follow Darwin's expression, but in the execution of which I have always been hampered by the insufficiency of the facilities for work that I had at my disposal."

This important passage shows that the works of Geoffroy Saint-Hilaire could without difficulty be interpreted in a Darwinian context. Serres was unable to take this turn. Cf. C. Dareste, *Recherche sur la production artificielle des monstruosités ou essai de tératologie expérimentale* (Paris: Reinwald, 1877); J. L. Fischer, "Camille Dareste," in P. Tort, ed., *Dictionnaire du darwinisme et de l'évolution* (Paris: Presses Universitaires de France, 1996), 1:767–769; J. Rostand, "Etienne Geoffroy Saint-Hilaire et la tératogenèse expérimentale," *Revue d'histoire des sciences* 17 (1964): 41–50.

5. "Our profound physiologist Lamarck has presented in his *Philosophie zoologique* considerations on the physical causes of life and the conditions it demands if it is to manifest itself. Adept at posing principles that he had drawn from calculated ideas of causality, he was less so in the choice of his particular proofs, in which he adduced a great number of facts that seemed to him to establish that the actions and habits of animals led in the long run to modifications in their organization."

6. "Paul has the desire and the means to procure for himself all the enjoyments of life: he is intelligent, inventive, and he has applied himself to investigate and to collect what he supposes must be necessary for him. He provisions his cellar with the best wines; he fills his woodshed with all the wood he will need to keep warm; he acts with the same discernment for all the other objects of his probable consumption. The qualities are well chosen, the objects cleverly arranged, and a wise order reigns everywhere. But when he has come so far, Paul stops. Of this wine he will not drink; from this wood he will not warm himself; of all the other pieces of his personal property he will make no use.—But, you tell me, your Paul is a madman.—I agree."

7. Cuvier's position was much weakened, due in particular to his antagonism with Arago, the second permanent secretary, as Goethe reports:

"At the Academy's session on July 19 we hear an echo of these differences, and now even the two permanent secretaries, Cuvier and Arago, come into conflict.

"Until then it was, as we have heard, the custom to report in each successive session only the rubrics of the previously presented numbers and thus to set everything aside.

"However, the other permanent secretary, Arago, this time makes an unexpected exception and presents in full detail the protestation that Cuvier had submitted. But the latter protests again at such innovations, which would necessarily take up a great deal of time, and at the same time complains about the incompleteness of the summary that has just been presented.

"Geoffroy Saint-Hilaire objects; examples from other institutes are cited, in which such proceedings are followed to good effect.

"Then there are further objections, and finally it is considered necessary to submit this matter to further consideration.

"At a session of October 11, Geoffroy reads an essay on the peculiar forms of the occiput of the crocodile and the *Teleosaurus;* here he reproaches M. Cuvier for neglect in the observation of these parts; the latter rises, much against his will, as he insists, but forced by these reproaches not to admit such remarks in silence. And this is a remarkable example of what great harm results when the conflict about higher views is given verbal expression in respect to particular details.

"Soon thereafter comes a session, which we wish to recall in the words of M. Geoffroy himself, as he reports in the *Gazette médicale* of October 23:

> Since the *Gazette médicale* and the other public papers have spread the news of the resumption of the old controversy between M. Cuvier and

me, people hurried to the session of the Academy to hear M. Cuvier in the development that he had promised to give on the auditory bulla of the crocodiles. The hall was full of the curious; consequently there were not only those zealous disciples animated by the spirit of those who frequented the gardens of Academe: one could distinguish manifestations of an Athenian crowd in the pit moved by quite different sentiments. This point, communicated to M. Cuvier, led him to postpone to another session the reading of his memoir. Armed with documents, I was ready to reply. However, I was glad of this solution. I prefer to an assault in the Academy the deposit that I am making here of the following summary, a summary that I had written in advance, and which, after improvisation became necessary, I had put back on my desk marked *ne varietur.*

"A year has already passed since these events, and my readers will be convinced by what has been said, that we have remained attentive to the consequences of so significant a scientific explosion, even after the great political one. But now, lest the point grow stale, we want only to explain that we believe we have noticed that scientific investigations in this field have long been treated by our neighbors with more freedom and in a more ingenious manner." *Goethes Werke* (Berlin and Leipzig: Bong, 1910), 36:311–313.

8. Sainte-Beuve reports this amusing fact.
9. In a letter written by Geoffroy Saint-Hilaire, and addressed to A. M. Tehler, member of the Academy of Sciences, we read:

   "My dear Colleague,

   "I cannot continue my investigations of the philosophical questions of paleontology; and although it cost me to write to you, and that it is with regret that I do it, I will confess to you that I lack the moral courage to continue; for the bruises to be submitted to are too numerous and incisive. Thus in the future I will keep my works in their portfolio. This is no longer a question accessible to the Academy. It would have to be admitted that G. Cuvier's scientific life has entered the public domain; but it is no longer only to the susceptibilities of his brother that one must reply. I have been forced to defend step by step the terms of my reply, when every license was permitted to the attack."
10. E. Geoffroy Saint-Hilaire, "Considerations on fossil bones, most of them unknown, discovered and observed in the basins of the Auvergne, accompanied by notes in which are explained the relations and the differences between two zoologies, that of the antediluvian epochs and that of the present world," *Rev. encycl.* 59 (1833): 76.
11. "Thus we have here the reasons for the special forms that differentiate the animals and the plants of the different ages of the earth, and especially the explanation of this fact, made more and more evident, to wit: that all kinds of special and different forms make use of the same base of materials, and are also invariably subject to similar laws for their respective arrangement. I flatter myself that I have arrived in this manner at

the demonstration that the two zoologies follow one another without a gap, without interruption, as engendered the one by the other, and as the result of modifications that have occurred under the action of time: they have each undergone the influence of the changes that every kind of ambient medium would have first undergone from age to age, changes that find themselves constituting the general movement of the universe."

12. "That is to insist that at bottom everything comes down to the methodological unfolding executed by the materials of which nature disposes, in a successive development capable of being included in the future as well as in the past, and managed definitively so that everything happens at a foreordained moment. It is in this sense that we can say of the birth of man that it was fixed from all eternity in the designs of Providence, in any case not to appear until the day that was foreseen, and as soon as the surrounding world that was to intervene with this result should have acquired all its stability and have been placed in position with respect to the elements needed so that man would be produced, that is to say, so that those elements, as forming its constituent and appropriate parts, would become capable of the association, of the coordination and of the harmonic relations indispensable for their coexistence."

13. In this passage Geoffroy Saint-Hilaire raises a problem that will come to the fore again only more than a hundred years later, that of the origin of life: "*It is quite certain that there was a moment when life did not yet exist on our planet, and another moment when it appeared. It is the passage between these two states that forms the great problem of natural philosophy today.*" In a letter that he will write later—July 13, 1838—to George Sand, he will explain his thought on this point: "*According to Cuvier, God has performed this immense miracle once and for all in order never to go back on it . . . That was quite simply the grossest stupidity that it was possible to propose to human credulity . . . I therefore adopted this thesis: God created materials predisposed for organization, by endowing them with all the virtual conditions to pass through all possible transformations according to the prescriptions of the unceasingly variable ambient media. Animal forms are thus unceasingly variable.*" For both of them, God was the creator, but not in the same manner. Cuvier thinks that God creates everything in an untouchable manner. Geoffroy Saint-Hilaire, for his part, thinks that God has given the materials and the conditions of transformation as a freedom bestowed on the living. Cf. F. Bourdier, "Le prophète Geoffroy Saint-Hilaire, George Sand et les saint-simoniens," *Histoire et nature* 1 (1973): 47–66.

14. On the death of his father, Pierre Leroux was not able to complete his studies at the Ecole polytechnique, for want of the financial means. He then became a typographic worker. Journalist in spirit, he never stopped starting new journals. He founded *Le Globe* in particular in 1824. Shortly after the July Revolution that journal became the official organ of Saint-Simonism. In 1840 he published *De l'Humanité,* in which he expounded "*the true definition of religion.*" He became the intellectual master of George Sand, who compared him to Rousseau. In 1848 he was elected to the Constituent Assembly, and then to the Legislative Assembly. He put forward a project

for a Saint-Simonist constitution. After the coup d'état of December 2, 1851, he fled to London and then to Jersey. He died during the Commune. History has remembered him as the creator of the word "*socialism*," used for the first time in the year 1833.

Geoffroy Saint-Hilaire cites in that note an extract from an article of Pierre Leroux, published in the quarterly review *De la Doctrine du progrès continu (On the doctrine of continual progress)*: *"By an admirable synchrony, all the contemporary discoveries reveal the constant change and the incessant creation of the universe; as they reveal the indefinite perfectibility of humanity. Here it is the school of anatomists who have long advanced philosophical intuition by proving the continuity of progress in the sequence of beings. There it is a nearly new science, geology, which steadily keeps renouncing its theories of cataclysms and of upheavals so as to explain the formation of our globe through a continuous development."* Cf. D. Bihoreau, *La Pensée politique et sociale en France sous le Second Empire* (Paris: Ellipses, 1995).

15. E. Geoffroy Saint-Hilaire, introduction to M. Buchez, *Introduction à la science de l'histoire, ou Science des développements de l'humanité* (Paris: Paulin, 1833). Report to the Academy, October 7, 1833. Philippe Buchez was a disciple of Saint-Simon and one of the first precursors of Christian socialism. He was president of the Constituent Assembly in 1848. He retired from political life under the Second Empire.
16. We find here a formulation that recalls the anthropic principle, recently formulated in the following manner: *"the presence of observers in the universe imposes constraints, not only on the age of the universe, from which time those observers are able to appear, but also on the totality of its characteristics and of the fundamental parameters of the physics which there unfolds." Cf. J. Barrow and F. Tipler,* L'homme et le cosmos (Paris: Imago, 1984); J. Demaret and Ch. Barbier, "Le Principe anthropique en cosmologie," *Revue des questions scientifiques* 152 (2), 152 (4) (1981): 181–222, 461–509.
17. Claude Henri de Rouvroy, count of Saint-Simon (1760–1825), is the distant cousin of the memorialist of Versailles. He gained his gold braid as a colonel during the American war of independence. At the Revolution he abandoned his title of count, and, speculating in state-owned property, he realized a considerable fortune which he squandered at the beginning of the Empire. For five years, from 1805 to 1810, he lived with an old servant. Thus he began to publish his books, which did not make an immediate impact. In 1814 his secretary was Augustin Thierry, and then in 1817 Auguste Comte. Saint-Simon's ideas lie at the origin of almost all the great currents of thought, like positivism and socialism.

    For Saint-Simon human society is like a gigantic living body, having its own physiology, that can perfect itself step by step in the course of time. Thus the study of such a body must pass through a positive phase of description.

    Saint-Simon forged a doctrine of production and believed in the virtue of elites. Thus he wished to reorganize the economy, the credit system and the spheres of decision, in order to ameliorate the moral and physical existence of the poorest class. He wished a change in property, going so far as to militate against inheritance.

Such an ideal was realized in certain concrete, important ways, like the first railroads, the plans for the Suez Canal (De Lesseps was a former Saint-Simonist), the personal and land banks of the brothers Péreire.

In 1831, Prosper Enfantin transformed Saint-Simonism into a religion. But many Saint-Simonists failed to accept this development, which would indeed come to grief. Among them we find Hippolyte Carnot, the son of Lazare, who bought the *Revue encyclopédique*, which would become—along with *Le Globe*, edited by Pierre Leroux,—the instrument of propaganda of the Saint-Simonists. See D. Bihoreau, *La Pensée politique et sociale en France au XIXe siècle* (Paris: Ellipse, 1995).

18. R. Bange and Ch. Bange, "Frédéric Gérard (1806–1857), un disciple de Lamarck et de Geoffroy Saint-Hilaire, théoricien de l'évolution," *Bull. Hist. Epistém. Sc. Vie*, 2, 1 (1995): 89–97.

19. H. de Balzac, *Les illusions perdus*, preface to the first edition, 1837, in Balzac, *La Comédie humaine* (Paris: Gallimard, Bibliothèque de la Pléiade, 1977).

20. After the controversy of 1830 Balzac oscillates in his perception of the opposition between the two zoologists. He poses as an admirer of Cuvier in *La Peau de chagrin* (1831) and *La Messe de l'athée* (1836). He finds himself once more in the clan of defenders of Geoffroy Saint-Hilaire in *Louis Lambert* (1833) and *Les Illusions perdues* (1837). In 1835, he thinks he will dedicate *Le Père Goriot* to George Sand. However, in the Furne edition of 1843, he dedicates it to Geoffroy Saint-Hilaire: "*To the great and illustrious Geoffroy Saint-Hilaire, as a testimony of admiration for his works and his genius.*"

It is evident that the composition of the preface to *La Comédie humaine* was of great importance for Balzac. It is there that he affirms his gratitude to Geoffroy Saint-Hilaire, whom he considers the victor in the controversy. It is there as well that the reference to Geoffroy Saint-Hilaire appears richest in meaning. Indeed, it is this text that will make the greatest impact on future generations:

> To begin with, the first idea of *The Human Comedy* was like a dream for me, one of those impossible projects that one caresses and then lets escape; a chimera that smiles, that shows its feminine countenance and at the same time uses its wings to remount to a fantastic heaven. But the chimera, like many chimeras, turns into reality; it has its commandments and its tyranny, to which it is necessary to submit.
>
> This idea comes from a comparison between Humanity and Animality.
>
> It would be a mistake to believe that the great quarrel that recently broke out between Cuvier and Geoffroy Saint-Hilaire rested on a scientific innovation. In other terms, the unity of composition was already occupying the greatest minds of the two preceding centuries. When we reread the quite extraordinary works of the mystical writers who were occupied with the sciences in their relations to the infinite, like Sweden-

borg, Saint-Martin, etc., and the writings of the finest intellects in natural history, like Leibniz, Buffon, Charles Bonnet, etc., we find in the monads of Leibniz, in the organic molecules of Buffon, in the vegetative force of Needham, in the nesting of similar parts in Charles Bonnet, who was bold enough to write in 1760, "The animal vegetates as the plant does"; we find, I say, the rudiments of the fair law of itself for itself on which the unity of composition rests. There is only one animal. The creator used only one and the same pattern for all organized beings. The animal is a principle that takes its external form, or, to speak more precisely, the differences of its form, in the media in which it is called on to develop. The Zoological Species result from these differences. The proclamation and the support of this system, in harmony, moreover, with the ideas that we forge for ourselves of divine power, will be the eternal honor of Geoffroy Saint-Hilaire, the conqueror of Cuvier on this point of high science, whose triumph was celebrated in the last article of the great Goethe.

Infused with this system well before the debates to which it gave rise, I saw that, in this context, Society resembled Nature. Does not Society produce from man, following the media in which his action is deployed, as many different men as there are varieties in zoology? The differences between a soldier, a laborer, an administrator, a lawyer, an idler, a scholar, a statesman, a tradesman, a sailor, a poet, a poor man, a priest, are, although more difficult to grasp, just as considerable as those that distinguish the wolf, the lion, the donkey, the crow, the shark, the seal, the sheep, etc. Thus there have existed, there will exist for all time, Social Species just as there are Zoological Species.

21. In 1836, George Sand composed a letter that she thought of integrating into a future novel. The text is edifying, entirely to the glory of Geoffroy Saint-Hilaire, but with a total incomprehension of his character:

Our epoch, wholly commercial, has not known how to look at the progress of the natural sciences; it has not understood that the great event was the struggle between Cuvier and his adversary. Cuvier, powerful, eloquent, indefatigable, decreed that the human mind must stop before certain mysteries; he wanted to make us return to fetishism, and to make us fear scholars, as our fathers once feared sorcerers.

His adversary, whose voice he tried to stifle, was an entirely different man, more religious than him in his respect for truth. In the great peace of the cloister, and in all serenity of mind, I read the illustrious pages of Geoffroy Saint-Hilaire; through their hard, strange, halting, obscure style, I discovered thoughts vast as the world, mysterious and grandiose cries of the soul, such as are found only at the dawn of religions. There shone through them an image of the Creation that satisfies more than others do the human mind's thirst for order and harmony; here is the universal chain of beings, its innumerable links united by equilibria and accords; here is lifeless nature joined to living nature; by insensible transitions, here is the stone that passes to the plant, the insect to the bird, the brute to man; man holds to all and all holds to God. There is no longer anything incomprehensible in the Creation; God is everywhere, intelligent in the way of man, a simple drop in his ocean of light.

This prophet guides us toward sanctuaries still veiled, where the unity of the creative principle is already physically demonstrated and where creation falls into place.

Strange vision, in which the materialism of the prophet does not appear! It is easy to understand George Sand's disappointment when, through his letters, she would come to understand his character.

F. Bourdier, "Le prophète Geoffroy Saint-Hilaire, George Sand et les saint-simonins," *Histoire et nature* 1 (1973): 47–66.

22. There is a reference to Geoffroy Saint-Hilaire in a recent work of Gilles Deleuze, about a commentary on a text of Freud: *"Moreover, Freud is much closer to Geoffroy Saint-Hilaire than to Darwin. Formulations of the type"* one does not become perverse, one remains it *"are copied from Geoffroy's formulations about monsters; and the two great concepts of fixation and regression come straight out of Geoffroy's teratology (*"arrest of development" *and* "retrogradation"*). Now Geoffroy's point of view excludes all evolution as direct transformation: there is only a hierarchy of types and of possible forms, in which beings stop more or less early, and regress more or less profoundly. It is the same with Freud: the combination of two kinds of impulse represent a whole hierarchy of figures, in whose order individuals stop earlier or later and regress more or less."* Unlike George Sand, Deleuze understands Geoffroy Saint-Hilaire perfectly. Cf. G. Deleuze, *Présentation de Sacher-Masoch* (Paris: Editions de Minuit, 1990).

Chapter Seven

1. F. de Chateaubriand, *Mémoires d'Outre-Tombe*, vols. 1 and 2 (1849; reprint, Paris: Gallimard, 1951).
2. We could cite the controversy on monstrosity between Lémery and Winslow, which began in 1724, the controversy on the status of disease between Piorry and Bousquet, which began in 1855, or, again, that on spontaneous generation between Pasteur and Pouchet, beginning in 1858. Looking back, it is not difficult, in these cases, to point out the one who was right.
3. Cuvier's behavior vis-à-vis Blainville is a good example in this regard. See E. L. Troussart, *Cuvier and Geoffroy Saint-Hilaire* (Paris: Mercure de France, 1909).

4. See, e.g., R. Owen, "Report on the Archetype and Homologies of the Vertebrate Skeleton," *Report of the British Association for the Advancement of Science* (Southampton Meeting), 169–340.
5. R. Thom, "La théorie des catastrophes et ses applications," in *Réflexions sur de nouvelles approches dans l'étude des systèmes* (Paris: Editions de l'ENSTA, 1975), 9–22; see also H. Le Guyader, "Zootype versus Blastula physiologique: à propos d'anatomie transcendante," in *Passion des formes. Dynamique qualitative, sémiophysique et intelligibilité, à René Thom* (Fontenay-Saint-Cloud: ENS, 1994), 547–567.
6. Social Darwinism, which developed chiefly in the course of the first half of the twentieth century, at the basis of eugenics and later of a certain sociobiology, will very quickly become a new program—criticizable and criticized—proposing an entirely different connection between the laws of nature and those of society. According to Becquemont, what is called social Darwinism is a certain form of sociology whose postulates are: (a) that since man is part of nature, the laws of human society are directly, or almost directly, those of the laws of nature; (b) that the laws of nature are the survival of the fittest, the struggle for existence, and the laws of heredity; (c) that it is necessary for the well-being of humanity to care for the proper functioning of those laws in society. Thus understood, social Darwinism can be defined, historically, as the branch of evolutionism that postulates a minimal gap, or none at all, between the laws of nature and social laws, both of them submitted to the survival of the fittest, and that considers the laws of nature to furnish directly a morality and a politics. See D. Becquemont, "Le darwinisme social," in P. Tort, ed., *Dictionnaire du darwinisme et de l'évolution* (Paris: Presses Universitaires de France, 1996).
7. C. Darwin, *The Descent of Man and Selection in Relation to Sex* (London: J. Murray, 1871).
8. The least one can say about this is that the vision of social Darwinism lacks the human warmth of Geoffroy Saint-Hilaire and the Saint-Simonists.
9. The universality of the concept of the cell was stated by Theodore Schwann (1810–1882) in 1838. See H. Le Guyader, *Théorie et histoire en biologie* (Paris: Vrin, 1988).
10. Pierre Belon (1517–1564), French physician and naturalist, is especially well known as an ornithologist. He gave his name to a duck, Belon's sheldrake. His magnum opus of 1555 is *L'Histoire de la nature des oiseaux, avec leurs descriptions et naïfs portraicts retirez du naturel* (*The history of the nature of birds, with their descriptions and unaffected portraits drawn from the natural state*).
11. At present thirty-two embranchements are enumerated in the animal world; nevertheless, many of them have only very few representatives. See C. Nielsen, *Animal Evolution: Interrelationships of the Living Phyla* (Oxford: Oxford University Press, 1995); R. C. and J. C. Brusca, *Invertebrates* (Sunderland, MA: Sinauer, 1990).
12. The medusae and the sea anemones belong to the group of Cnidaria, which are simple organisms with radial symmetry, made of two layers of cells; their digestive tube has only one opening. Nevertheless, they can form large colonies—like the corals of the coral reefs—or giant medusae, like the Physalia.

13. The organisms with three germ layers have a digestive tube with two orifices—apart from a few exceptions, which are easily explicable; they are basically bilaterally symmetrical; that is why they are called Bilateria.
14. The concept of biodiversity is difficult to summarize in a few words. We may give here the definition proposed by the report on global strategy on biodiversity published by the World Resources Institute, the United Nations Program for the Environment, the United Nations Program for Food and Agriculture, and the United Nations Program for Education, Science, and Culture (UNESCO): "Biodiversity is the totality of the genes, species and ecosystems of a region." Three major levels are thus determined. Genetic diversity denotes the diversity of genes within populations and species; specific diversity, that of species in a region; and finally, ecosystemic diversity, that of ecosystems. It is clear that at the time of Geoffroy Saint-Hilaire, there was interest only in the diversity of species.
15. William Bateson also had considerable importance at the beginning of the century. He was one of the leaders of the new discipline of genetics, which appeared to be opposing a Darwinian reading of evolution. See W. Bateson, *Materials for the Study of Variation* (London: Macmillan, 1894).
16. The rediscovery of Mendel's laws is due independently, in 1900, to Carl Correns (1864–1933), Erich von Tschermak (1871–1962), and Hugo de Vries (1848–1933). The chromosomal mechanism of heredity was studied by the school of Thomas Hunt Morgan (1866–1945), thanks to the utilization of mutants of the fruit fly *Drosophila*. Cf. M. Morange, *History of Molecular Biology* (Cambridge: Harvard University Press, 1998).
17. S. F. Gilbert, *Developmental Biology* (Sunderland, MA: Sinauer, 1994); P. A. Lawrence, *The Making of a Fly* (Oxford: Blackwell Scientific Publications, 1992).
18. If Edward Lewis arrived at the concept of the homeotic gene, it is to Christiane Nüsslein-Vollhard and Eric Wieschaus that we owe the description of the whole of the cascade of developmental genes that intervene before the expression of the homeotic genes. All three received the Nobel Prize in Physiology or Medicine in 1995 for the discovery of the precocious genetic control of embryonic development. See W. McGinnis and M. Kuziora, "Les gènes de développement," *Pour la science*, L'évolution, dossier hors-série, January 1997; J. Deutsch et al., "Prix Nobel de médecine 1995," *Médicine/Sciences* 10 (1995): 1625–1628.
19. The homeobox is a sequence of 183 base pairs, coding a protein sequence of 161 amino acids, called the homeodomain. The role of the homeodomain is now known. A protein that is provided there can be linked to a molecule of DNA; this capacity is that of genes that regulate the expression of other genes. In fact, the homeotic genes are regulatory genes governing genetic expression in the course of development.
20. The Cambrian layer of the Burgess shale is the chief level at which these fossils have been found. Arthropods, or at least a chordate, have been described, if we accept the classic interpretation of the fossil *Pikaïa*. More recently a bed with an equivalent fauna has been found in China, in the Yunan, at Chenjiang. Greenland also presents

beds in the course of study. See D. E. G. Briggs et al., *The Fossils of the Burgess Shale* (Washington: Smithsonian Institution, 1994); S. Conway Morris, "The Fossil Record and the Early Evolution of the Metazoa," *Nature* 361 (1993): 219–225; S. J. Gould, *Wonderful Life* (New York: Norton, 1989).

21. Research has been carried out in nonsegmented worms, in particular in the nematode *Caenorhabditis elegans*. Homologous genes have been found there; here, in contradiction to what had first been published, they are interpreted as grouped in a complex of four genes, as in the insects or the vertebrates. Genes of the family antennapedia have also been investigated in an annelid, a leech. Everything leads us to believe that we are in the presence of a gene complex. The planarians are provided with homeobox genes. Further, homeobox genes have been found in diploblasts, such as the hydra *Chlorohydra viridissima*. But the diploblasts have only a single labial type gene.

    Thus it seems reasonable to postulate that the distribution into organized complexes exists in the triploblasts before the separation of insects and vertebrates. Evolution has thus proceeded in three ways: (1) duplication of genes in the interior of a complex; (2) gradual further specialization of these genes, but also of the regulatory circuits associated with them; (3) duplication of whole complexes, especially in the vertebrates. See B. L. Aerne et al., "Life Stage and Tissue-Specific Expression of the Homeobox Gene cnox1-Pc of the Hydrozoan *Podocoryne carnea*," *Developmental Biology* 169 (1995): 547–556; E. Pennisi and W. Roush, "Developing a New View of Evolution," *Science* 227 (1997): 34–37; F. H. Ruddle et al., "Evolution of Hox Genes," *Annual Review of Genetics* 28 (1994): 423–442; F. R. Schubert et al., "The Antennapedia-Type Homeobox Genes Have Evolved from Three Precursors Early in Metazoan Evolution," *Proceedings of the National Academy of Sciences U.S.A.* 90 (1993): 143–147; B. B. Wang et al., "A Homeobox Gene Cluster Patterns the Anteroposterior Body Axis of *C. elegans*," *Cell* 74 (1993): 29–42.

22. In reality, this definition should be confined to the triploblasts or Bilateria. See J. M. W. Slack et al., "The Zootype and the Phylotypic Stage," *Nature* 361 (1993): 490–492.

23. For each phylum at present being considered, the zootype is best found at the moment of a particular stage of embryology called the phylotypic stage—the "Körpergrundgestalt" of Seidel. It is the stage of development at which the general characteristics of the animal's plan of organization can most clearly be detected. To put it more precisely, it is the stage at which, for example, the principal parts of the body are detectable in their final disposition, by condensations of undifferentiated cells; this could equally well be the stage corresponding to the end of morphogenetic movements, or, again, the stage at which the representatives of a phylum present the maximum degree of similarity.

24. See chapter 2, n. 15.

25. The antiquity of the zootype could be brought into relation with a capital evolutionary fact, that of the probable explosion of life in the Cambrian, particularly

clearly evidenced by the study of the fossils of the Burgess shale, and strongly suspected by molecular phylogeny. Such a rapid diversification of forms demands explanation, and it has found two kinds—which do not exclude one another. One can postulate external conditions that were decisive, such as, for example, the increase of the level of oxygen in the atmosphere and in the water, or again the presence of many empty ecological niches. One can equally well stress internal changes, like the appearance of the mesoderm, and then of the coelom. Following this sort of idea, the appearance of an imposing battery of genes ready to be coopted for development would have been a key factor, opening the possibility of many structural complexifications through a process of coding along the anterior-posterior axis of the organism. The appearance of new plans of organization could then have taken place very rapidly, following a diversificatory explosion. See S. A. Bowring et al., "Calibration Rates of Early Cambrian Evolution," *Science* 261 (1993): 1293–1298; D. E. G. Briggs et al., *The Fossils of the Burgess Shale* (Washington: Smithsonian Institution Press, 1994); S. J. Gould, *Wonderful Life* (New York: Norton, 1989); D. K. Jacobs, "Selector Genes and the Cambrian Radiation of Bilateria," *Proceedings of the National Academy of Sciences U.S.A.* 87 (1990): 4406–4410; S. Conway Morris, "The Fossil Record and the Early Evolution of the Metazoa," *Nature* 361 (1993): 219–225.

26. The gene *short gastrulation* (*sog*) represses the gene *decapentaplegic* (*dpp*) in drosophila. The gene *brown morphogenetic protein 4* (*BMP 4*) represses the gene *chordin* in xenopus. The dorsoventral determinations are inverted. See E. M. De Robertis and Y. Sasai, "A Common Plan for Dorsoventral Patterning in Bilateria," *Nature* 380 (1996): 37–40; E. L. Ferguson, "Conversation of Dorsal-Ventral Patterning in Arthropods and Chordates," *Current Opinions in Genetics and Development* 6 (1996): 424–431; S. A. Holley, "A Conserved System for Dorsal-Ventral Patterning in Insects and Vertebrates Involving *sog* and *chordin*," *Nature* 376 (1995): 249–253; C. M. Jones and J. C. Smith, "Revolving Vertebrates," *Current Biology* 5 (6) (1995): 574–576; T. Lacalli, "Dorsoventral Axis Inversion: A Phylogenetic Perspective," *Bioessays* 18 (1996): 251–254; K. Nühler-Jumg and D. Arendt, "Is Ventral in Insects Dorsal in Vertebrates?" *Roux Arch. Dev. Biol.* 203 (1994): 357–366.
27. The vertebrates are deuterostomes, a word of Greek origin that means "second mouth."
28. S. J. Gould, "Geoffroy and the Homeobox," *Natural History*, November 1985, 12–23.
29. C. M. Jones and J. C. Smith, "Revolving Vertebrates," *Current Biology* 5 (6) (1995): 574–576.
30. E. M. De Robertis and Y. Sasai, " A common plan for dorso-ventral patterning in Bilateria," *Nature* 380 (1996): 37–40. The Bilateria are organisms with three germ layers. We have seen that the comparison of insects and vertebrates was expressed at length by Geoffroy Saint-Hilaire in three memoirs of 1819 and 1820. The text of 1822, the only one cited by the English-language authors, concerns more particularly the dorsoventral orientation.

INDEX

Italicized page numbers refer to figures.